Praise for

My Beloved Brontosaurus

One of *Publishers Weekly*'s Top Ten
Spring 2013 Science Books

Selected by Apple's iBookstore as
One of the Best Books of April 2013

"Making scientific concepts accessible using a playful voice, in the manner of Neil deGrasse Tyson and Malcolm Gladwell, Switek guides readers through the ever-shifting world of conventional dino-wisdom, from why they became the dominant life form of their time, to what they looked like outside those giant bones, and even covering how they might have had sex." —*Salt Lake City Weekly*

"Switek has a knack for finding fascinating specifics and present-ing them in engaging ways. He excels at relating fossil finds to their once-living counterparts, giving these animals an awesome sense of reality. Even readers whose younger days of dinosaur frenzy are long extinct will find *My Beloved Brontosaurus* a fasci-nating read." —*Shelf Awareness* (starred review)

"*My Beloved Brontosaurus* is, in many ways, science writer Brian Switek's love letter to his favorite animal . . . But the book is more than a personal ode to fossils . . . [Switek] is searching no longer for preserved eggs that might one day hatch dino-babies but rather for an understanding of what life in the age of dinosaurs was like." —Maggie Fazeli Fard, *The Washington Post*

"Switek geeks out gloriously on everything from the truth about *Jurassic Park* to the ugliest roadside dinosaurs he has ever seen. He's a friendly guide to the latest in dinosaur science, whizzing through journal article after journal article on how cross sections of dinosaur bone can tell us about their physiology, or how fossil olfactory lobes can reveal their sense of smell through time . . . Much has been written about the 'dinosaur revolution' of recent decades, in which the idea of sluggish, passive reptiles transformed into more active and engaging creatures. Switek takes the science a step farther and into the twenty-first century."

—Alexandra Witze, *The Dallas Morning News*

"Switek passionately and playfully explores scientists' evolving perception of the wild, wonderful dinosaur world, emphasizing at every turn the dynamic nature of their field despite its now inanimate subjects . . . Switek intersperses his rich, well-researched scientific and historical discussions with personal anecdotes and cultural signposts, weaving together a narrative that reveals the current state of the field as well as some of the wrong turns along the way." —Rachel Bernstein, *Science*

"Switek earned fame as the unabashed dinosaur advocate behind the fossil-focused blog *Laelaps*. He applies that same blend of boyish exuberance and serious science to this exploration of paleontology's roots, revisions, and future course . . . His zeal is infectious . . . Switek fleshes out the monstrous skeletons that we all remember from childhood museum field trips with meaty new findings about their anatomy and behavior." —*Discover*

"Switek's writing is crisp and clean, and he knows his dinosaurs . . . [He] does a good job of keeping up with the latest refinements in dinosaur science. Crucially, he shares his enthusiasm well, writing about the fun, the weird and the wonderful without the tall tales of the explorers of old." —Jeff Hecht, *New Scientist*

"Fortunately for us, Brian Switek has continued to channel the enthusiasms of his own inner five-year-old. In his zany, sometimes mind-blowing romp through the new science of old bones . . . you too can nerd out anew . . . The discoveries Switek shares fill the

old world with new wonders . . . Switek rekindles that childhood amazement at how vast the dinosaur world really was."
—Tess Taylor, *Barnes & Noble Review*

"[Switek] deftly brings the concerns of dinosaur fanatics regarding the dinosaurian public image to the popular culture . . . It's always a treat to read someone so captivated by the romance of natural history . . . Switek's message is that dinosaurs are a relevant, vital field of study because the illumination extends to broader issues of evolution and the fate of our world. In the same way, *My Beloved Brontosaurus* uses the engaging topic of dinosaur lives as a way to celebrate the ongoing exploration of science at large."
—David Orr, *Love in the Time of Chasmosaurs*

"A lovely, bite-size bit of pop science . . . A wonderfully smart book that can be easily digested with only a bare minimum of previous dinosaurs knowledge."
—*The Pretty Good Gatsby*

"*My Beloved Brontosaurus* walks readers through the new science of dinosaurs, and most importantly how scientists know these things to be true. It's a fascinating read."
—Chad Orzel, *Uncertain Principles*

"Cleverly written, well researched and sprinkled with lots of tidbits of information gleaned from dinosaur sites that include Utah's wealth of material, Brian Switek's *My Beloved Brontosaurus: On the Road with Old Bones, New Science, and Our Favorite Dinosaurs* is a jewel."
—Sharon Haddock, *Deseret News* (Salt Lake City)

"Fascinating . . . Switek's scope of knowledge is awesome . . . This is in the classic 'news you can use' for nerds genre."
—Razib Khan, *Gene Expression*

"If you are itching to learn more about dinosaurs than just the bare bones, so to speak, then Brian Switek's *My Beloved Brontosaurus* is a must-read . . . In addition to being packed with fresh research, this book is just flat-out fun to read . . . Perhaps Switek's greatest triumph with *My Beloved Brontosaurus* is proving beyond a shadow of a doubt that dinosaurs are not some esoteric, irrelevant subject, but are deeply connected to our own human story."

—Becky Ferreira, *The Mindhut*

"[*My Beloved Brontosaurus*] is a lifeline for the dinosaur enthusiast— an entertaining guide to the latest science of dinosaurs."

—Allison Bohac, *Science News*

"Switek's chatty, informative cross-country adventure is the perfect read for catching up on the latest, most fascinating dino science."

—*Mental Floss*

"[*My Beloved Brontosaurus* is] a wonderful overview of current research and knowledge of dinosaurs, for both lifelong dino-lovers and people who need an introduction to the prehistoric beasts . . . Informal, often humorous (in the tradition of great nonfiction writers like Mary Roach and Bill Bryson), without sacrificing scientific detail. The result is both readable and highly entertaining . . . An excellent field guide to the real dinosaurs that walked the planet."

—Matthew Francis, *DoubleXScience*

"You won't find a better guide to paleontology than Brian Switek, a fossil-crazed writer whose clear-eyed skepticism never dampens his boyish enthusiasm. And why should it? Dinosaurs, Switek shows convincingly, need no hype to blow your mind. The man is fearless. This book is splendid."

—David Dobbs, author of *My Mother's Lover* and *Reef Madness*

"Writing with unaffected ardor, Switek will resonate with readers fascinated by dinosaurs."

—*Booklist*

"There are so many dinosaur books—but *My Beloved Brontosaurus* is something special. Brian Switek, a self-confessed dinosaur fanatic, imparts his enthusiasm in a lively, thoroughly entertaining and carefully documented way. It is a joy to read such a well-

researched and contemporary account of dinosaurs written for non-specialists . . . This hard-to-put-down volume includes all manner of tidbits, from amorous penguins to alligators dying of sunstroke. Well written and creatively structured, this beguiling work is enriched with personal experiences. And when readers reach the end of the book they will feel a sense of loss, like the demise of the incredible creatures captured between its covers."
—Christopher McGowan, Curator Emeritus, Vertebrate Paleontology, Royal Ontario Museum; author of *The Dragon Seekers*

"An enthusiastic account of the history, description, discoveries, ongoing controversies and inaccurate media obsession with these popular but extinct creatures . . . A genuinely informative introduction to [Switek's] favorite subject." *Kirkus Reviews*

"Brian Switek has a true passion for the giants of the ancient past. Whether you are new to the world of the 'fearfully great lizards' or are a lapsed dinosaur fanatic, this book will help you understand how paleontologists bring fossils to life."
—Thomas R. Holtz, Jr., Senior Lecturer, Department of Geology, University of Maryland; author of *Dinosaurs: The Most Complete, Up-to-Date Encyclopedia for Dinosaur Lovers of All Ages*

"With *My Beloved Brontosaurus*, Brian Switek reaffirms his status as one of our premier gifted young science writers. It's an exciting time for dinosaur research, and Brian is the best guide I know."
—Kevin Padian, University of California Museum of Paleontology, Berkeley

"Readers will readily agree when Switek concludes that 'dinosaurs are better than ever.'" —*Publishers Weekly*

"Brian Switek is my favorite dinosaur tour guide in the world—and his book is smart, funny, lyrical, and can't-put-it-down readable."
—Deborah Blum, bestselling author of *The Poisoner's Handbook*

Brian Switek
My Beloved Brontosaurus

Brian Switek is an online columnist for *National Geographic* and is the author of *Written in Stone*. He has written for *Smithsonian, Wired, Slate, The Wall Street Journal, Nature, Scientific American,* and other publications. His examinations of fossil discoveries have been featured by the BBC and NPR. He lives in Salt Lake City, Utah. Follow him on Twitter at @Laelaps. Visit his website at www.brianswitek.com.

My Beloved Brontosaurus

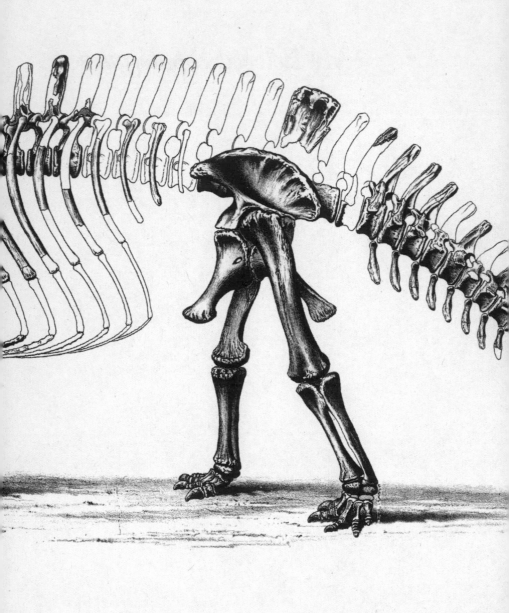

My Beloved Brontosaurus

On the Road with Old Bones, New Science,

and Our Favorite Dinosaurs

Brian Switek

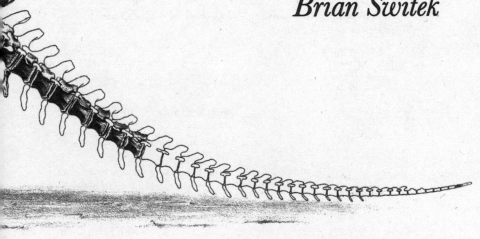

Scientific American / Farrar, Straus and Giroux

New York

Scientific American / Farrar, Straus and Giroux
18 West 18th Street, New York 10011

Copyright © 2013 by Brian Switek
Printed in the United States of America
Published in 2013 by Scientific American/Farrar, Straus and Giroux
First paperback edition, 2014

An excerpt from *My Beloved Brontosaurus* originally appeared,
in slightly different form, in *Scientific American*.

The Library of Congress has cataloged the hardcover edition as follows:
Switek, Brian.
My beloved Brontosaurus : on the road with old bones, new science, and
our favorite dinosaurs / Brian Switek. — First edition.
    p. cm.
ISBN 978-0-374-13506-5 (hardback)
    1. Apatosaurus—Miscellanea.   2. Paleontology—Miscellanea.
3. Popular culture.   I. Title.

QE862.S3 S94 2013
567.913'8—dc23

2012034530

Paperback ISBN: 978-0-374-53426-4

Designed by Jonathan D. Lippincott

www.fsgbooks.com • books.scientificamerican.com
www.twitter.com/fsgbooks • www.facebook.com/fsgbooks

Scientific American is a trademark of Scientific American, Inc.
Used with permission.

1 2 3 4 5 6 7 8 9 10

To Scicurious
A dear friend who remembers the old
"thunder lizard" as fondly as I do

Contents

My Beloved Brontosaurus

The author as a young dinosaur fan. (Photograph courtesy
Barbara Switek)

Prologue: My Dinosaur Life

I was once a dinosaur. A *Stegosaurus*, to be exact. My uncomfortably snug green jumpsuit, with floppy fabric plates sewn along the back, was far from scientifically accurate, but that wasn't important. I had the dinosaur spirit. That's what counted.

I had been tapped as one of the main players in an *Allosaurus*-versus-*Stegosaurus* battle set for my preschool's Dinosaur Night. It was yet another chance to coerce my parents to let me frolic among dinosaurs. My teachers had hidden little plastic dinosaurs in a shallow sandbox dig site, and everyone got a box of bland dinosaur cereal at the end of the night, but if there was any educational subtext, I can't remember it. That didn't matter to me at the time. Who needs a reason to play with dinosaurs when you're a five-year-old prehistory fanatic?

I was ready to roar, stomp, and swing my spiky tail at the *Allosaurus* kid in a Jurassic death match when I noticed that he was dressed in an identical costume. My opponent looked nothing like the nimble hypercarnivore with steel-trap jaws he was supposed to be. My teachers hadn't done their homework. And I didn't agree with my scripted defeat at the claws of *Allosaurus*, either. At the moment of my mock death, when I was supposed to fall and bare my reptilian throat to my attacker, I decided to break character and try to convince my audience that *Stegosaurus* was really the superior dinosaur. *Allosaurus* was fierce and fast, I explained, but those attributes would have been of little use against the prominent plates and bone-piercing tail spikes of *Stegosaurus*.

Alas, the assembled adults didn't appreciate my impromptu dinosaurology lesson. I was hoping the grown-ups would sagely nod their heads in agreement and recommend me for a post at the American Museum of Natural History in New York City. But they just laughed.

I didn't shake my fist and scream "Fools! I'll show you all!" as I felt in my heart a true scientist should do, but I didn't give up on dinosaurs, either. I nurtured my dinomania with documentaries, delighted in the dino-themed B movies I brought home from the video store, and tore up my grandparents' backyard in my search of a perfect *Triceratops* nest. Never mind that the classic three-horned dinosaur never roamed central New Jersey, or that the few dinosaur fossils found in the state were mostly scraps of skeletons that had been washed out into the Cretaceous Atlantic. My fossil hunter's intuition told me there just *had* to be a dinosaur underneath the topsoil, and I kept on excavating my pit. That is, until I got the hatchet out of my grandfather's toolshed and tried to cut down a sapling that was in my way. My parents bolted out of the house and put a stop to my excavation. Apparently I hadn't filled out the proper permits before I started my dig.

That's not to say that my parents didn't otherwise support my fossil infatuation. They encouraged my paleontological dreams, and I will always cherish the memory of them defending my book choices when my elementary school's librarian complained that I was checking out too many dinosaur titles that were supposedly above my reading level. My brain ached with the need to know everything there was to know about dinosaurs. Every new dinosaur name I learned became a scientific incarnation, a magic word that immediately conjured up terrible, marvelous, scaly monsters in my imagination.

Two and a half decades later, my wife now copes with the dinosauriana that is rapidly radiating out from my desk and creeping into every room of our Salt Lake City apartment. My dinosaur dreams factored into our decision to move here, too. When people ask me why in the world I would want to move to Utah, a place whose Mormon legacy includes maddeningly conservative politics and "near beer," my answer is very simple: "For the dinosaurs." With apologies to Horace Greeley, my rationale for coming to Utah was "Go West, young man, and grow up with the dinosaurs." The Beehive State is home to some of the richest dinosaur-bearing formations anywhere, all laid out in arid, colorful badlands. And while other couples might go back and forth about whether they can afford a new couch or television, I tend to spend hours trying to wear down my wife's financial resolve to let me bring home essential items like a full-size cast of an *Apatosaurus* skull from an estate sale. (The plaster sauropod head now sits triumphantly atop one of my paleontology-dedicated bookshelves.)

I only get to join the search for more dinosaurs when the weather allows, though. After October, the weather is too cold to go prospecting and the ground is too hard to safely extricate fossils. To pass the time I spend my winters writing about the rushing flow of new paleo papers, anxiously awaiting spring. Every new field season brings new possibilities. Despite what you might expect from more than a century of fossil hunting in the American West, there are still many dinosaurs left to discover. I haven't found that *Triceratops* nest just yet, but now I live in a place one step closer to my dreams, where Earth's history is thrust up and exposed in beautiful, fossil-rich swaths.

Only, dinosaurs aren't supposed to be part of adulthood.

Paleontologists and volunteer dinosaur diggers are often seen as overgrown children who somehow found a way to make a profession out of playing in the dirt and dreaming of snaggletoothed horrors treading through primordial ooze. American kids are expected to go through a "dinosaur phase," but we're due to give that up once we discover team sports and kissing under the high school bleachers. (My natural awkwardness prevented me from doing either of those things.) An awakening to cathartic music, the sheer terror of dating, and a concentrated push to narrow down dreams into viable career options are scheduled to take over and sweep out all the childhood clutter. Let's face it: dinosaurs have been culturally demarcated as kitschy kid stuff—triggers for nostalgia and ironic whimsy, but not a subject to take seriously.

At least not until the former dinosaur fans have kids of their own and take their broods to see the real-life monsters that stalk museum halls. The dinosaurs they grew up with are gone, replaced by creatures that look entirely different and sometimes don't even carry the same names. The dinosaurs we meet as children don't stay around long—science is always tweaking and refining them, giving us a jolt when we're expecting comforting memories.

I experienced a similar shock when my once-girlfriend Ellen took me to see the American Museum of Natural History's dinosaurs on New Year's Day 2003. I hadn't visited the halls since I was a kid, and in the intervening years the museum had renovated its fossil exhibits. The skeletons that had so inspired me as a youngster had received a fantastic makeover.

Tyrannosaurus rex, as I first met her, reared back in a Godzilla pose, fang-lined jaws held high and tail dragging on the ground. The *Stegosaurus* I encountered on the same visit looked like a mound of plates and spikes, while the low-slung *"Brontosaurus"* stood there stupidly, looking out of place on dry land rather than in a fetid, weed-choked pool. The dinosaurs I grew up with lived in a humid, slow-motion nightmare of teeth, claws, and horns. All had been replaced by unfamiliar visions of Mesozoic life—dinosaurs that stood tall and frozen in midstride as if their flesh had suddenly fallen off as they sauntered through their world. The new dinosaurs, poised in active skeletal snapshots, were strangers to me.

The fossilized vestiges of dinosaurs—their actual bones—remain unchanged. Individually, they appear to be static monuments of an era far beyond the reach of history. But since the time I first set eyes on the dinosaurs, paleontologists have been developing ever-more-refined techniques for gleaning information about prehistoric lives from those remains. A sauropod femur or a hadrosaur skull is not just a petrified lump meant only to be mounted and then left to collect dust. Every dinosaur fossil contains clues about that animal's life, evolution, and sometimes even death. Understanding a dinosaur doesn't stop with reassembling the puzzle of their bones—that's just the start of paleobiological reinvention.

What earlier generations could only speculate about, we can now begin to investigate. Everything from dinosaur sex lives to that most persistent of mysteries—what colors they were—is coming to the forefront. And the more we learn, the more peculiar and spectacular dinosaurs become. The *Tyrannosaurus* that I first met has been ripped to shreds by a much more active and fascinating carnivore—a muscular apex predator with a spine held parallel to the ground, a hot-running metabolism, and a fine coat of dinofuzz, betraying the tyrant's distant relationship to modern birds. *Stegosaurus* and all the other classics have been rearticulated and revitalized, too—moved out of the prehistoric bayous and cast in colors as vibrant as their natural history.

But it takes time for discoveries to reach the public, and even then, the science of how we know so much about dinosaur biology is often obscured. Museum halls and documentaries may present the products of paleontology—fleeting glimpses of dinosaurs inside and out—but rarely do they explain *why* dinosaurs have changed so much. Those secrets are revealed only in symposia and technical papers beyond the reach of casual dinosaur fans. It's impossible for even the acutely dinosaur-attuned to keep up with the pace of discovery. Change overtakes our understanding so rapidly that even sparkling museum displays are at least partly out of date by the time they open to the public. From estimates of dinosaurs' total length to the positions of their nostrils, paleontologists are constantly revising and arguing over what the beasts were really like. In all of this, skeletal reconstructions and life restorations are often taken as the last word when they are really working hypotheses, open to change and revision at the drop of a journal article. And, oddly enough, paleontology is one of the few areas of science where it's fashionable to distrust changes in our understanding, a love of tradition leading prehistory fans to mock feathery tyrannosaurs as giant chickens and weep over the loss of a favorite dinosaur species thanks to taxonomic arcana.

Of all the creatures to be caught between science and public affection, it is *"Brontosaurus"*—the iconic dinosaur that suffered a second extinction at the hands of research—that stays with me. It is "betwixt and between," a dinosaur properly called *Apatosaurus* and yet known and cherished by its old name. *"Brontosaurus"* is an unofficial nickname we're not supposed to use anymore, but we can't let go of it.

A bulky hill of animate flesh, *"Brontosaurus"* was the epitome of what it was to be a dinosaur. I remember her fondly. The long-necked giant was my introduction to how magnificent dinosaurs were, but she evaporated into the scientific ether just as soon as I met her. Today, *"Brontosaurus"* lives on only as a memory. But I cherish that memory, and I'm not alone. *"Brontosaurus"* is an icon that embodied the lifestyles of the big and scaly. To hear that the dinosaur didn't exist felt less like a technical mistake than a betrayal.

"Brontosaurus" is this book's mascot. There is no better symbol of the tension between the actual animals paleontologists investi-

gate and the pop-culture images of these behemoths—visions of prehistoric life that take on a life of their own and can dig their claws into our imagination as tenaciously as any *Velociraptor.* That's the danger, and fun, inherent in the process of discovery. In order to understand dinosaurs, we need to resurrect them in fiberglass, steel, paint, and computer-generated models. Inevitably, these older visions battle with updated versions of themselves. Scientific discovery catalyzes violent competition between what we thought we knew and what we currently understand. *"Brontosaurus"* is the most famous casualty of these perpetual skirmishes, but she is not only that. She is a familiar milestone that we can use to measure just how much science has changed dinosaurs. We may have lost a dear dinosaur, but the same process that destroyed the titan has revealed clues about prehistoric lives that we never expected to find. With *"Brontosaurus"* as our traveling companion, let's catch up with some old friends and see what secrets they've begun to teach us about evolution, extinction, and survival.

One

Dragons of the Prime

"*Brontosaurus*" will always be special to me. For my younger self, especially, the shuffling, swamp-dwelling hulk was an icon of everything dinosaurs were supposed to be—big, scaly, and, most of all, so thoroughly bizarre that they could only have belonged to a primeval past. And, though dead for over 150 million years, "*Brontosaurus*" lived on in my imagination. From the time I was a toddler, I desperately wanted to meet the gigantic herbivore. In my preschool scribblings, I included a pet "*Brontosaurus*" in crayoned portraits of my family. I kept it reasonable. I knew we could never afford an eighty-foot dinosaur, so I went with a Bronto roughly the size of a Great Dane. She was big enough to let me ride on her back, but small enough that my parents wouldn't go poor providing appropriate forage for my friend.

Resuscitating the dinosaur in Crayola colors barely even touched the depths of my dinomania. When my parents drove my siblings and me to Disney World for the first time, I so fiercely harried them about seeing the animatronic "*Brontosaurus*," *Stegosaurus*, and kin at the Exxon-sponsored Universe of Energy attraction that Mom and Dad didn't even unpack the car before putting us on the right bus to see the dinosaurs. Forget Mickey and Minnie. The jerking, wailing robotic dinosaurs were at the top of my list. And while I would later curse being stuck in the mind-numbingly mundane confines of central New Jersey, my captivity in the suburban sprawl carried at least one advantage. There was scarcely a better place for a young dinosaur fan than the nearby American Museum of Natural History, just over the river in New York City. That's where I first met my favorite dinosaur.

The museum no longer looks like it did when my parents guided my younger self up to the fourth-floor dinosaur halls in 1988. Today, the white walls, high ceilings, and ample illumination make the skeletons of *Tyrannosaurus*, *Edmontosaurus*, *Triceratops*, and other dinosaur celebrities stand out in sharp contrast from their surroundings. This open, airy vibe was created by a renovation project in the mid-1990s to adjust the prehistoric stars in accordance with new discoveries. Arranged in an evolutionary rank and file, the revised halls are a testament to how much dinosaurs have changed since nineteenth-century naturalists first recognized them. The AMNH dinosaurs stand alert, skeletal heads and tails at attention as if they're scanning a vanished landscape for food, friend, or foe.

During my early twenties, when I had the freedom to visit whenever I pleased, I took any chance I could get to wander among these skeletons and imagine flesh on their bones. And, as I strolled through those halls, the floors scuffed by the feet of so many youngsters on their first trips into the presence of dinosaurs, what I missed most was the dim, dusty Jurassic Dinosaur Hall that I encountered so many years before. The old dinosaurs were horribly wrong when I viewed them back in the 1980s—awkward aberrations ultimately sent to the scientific trash heap—but that doesn't diminish my memory of seeing them for the first time. Way back then, in the forbidding gloom of the hall, my imagination gave the bones a thin cast of vitality. The skeletons felt less like perished monuments to paleontology and more like bony scaffolding waiting to be connected by sinew and wrapped in scaly hides. My young mind didn't see dead dinosaurs, but the osteological architecture of creatures that might walk again.

•

I was so consumed by the idea during my first trip to the AMNH that I can hardly remember my parents being there. Standing beneath the prehistoric skeletons, I was entranced. I couldn't take my eyes off the museum's *"Brontosaurus,"* with her neck stretched low, tipped with a moronic blunt skull full of spoon-shaped teeth. I was in the court of the queen of all sauropods—the long-necked, heavy-bodied dinosaurs that were the largest creatures ever to walk the Earth. After all, as my schoolbooks told me, *"Brontosaurus"* was so massive that her name meant "thunder lizard." When she walked, it must have sounded like a storm rolling across the Jurassic landscape. I imagined that sound as I admired her skeleton. She seemed poised to step off the platform, duck out the exit, and plod right down to the foliage along Central Park West. In the intense quiet of that moment, I could have sworn that I heard the ethereal remnant of the dinosaur's breathing. In a place with so many prehistoric bones, there had to be ghosts.

Yes, the old mounts of *Tyrannosaurus* and other dinosaurs were impressive, too. But they didn't stick with me quite like the *"Brontosaurus."* I couldn't help but wonder what it would have been like to catch a glimpse of the dinosaur trundling down my street, picking succulent leaves from the oaks of my neighbors' lawns. I drew sluggish brontosaurs in my school art portfolios, made my plastic sauropod models bask in an improvised mud puddle I created in the driveway's storm drain, and dreamed of some far-off swamp where the dinosaur might still sun itself, enjoying a reprieve from extinction.

And then I heard the bad news.

"Brontosaurus" was dead to begin with. My favorite dinosaur wasn't real, but only a misconstrued amalgamation that had been borne and slaughtered by science. The dinosaur's true name was *Apatosaurus*—a creature that paleontologists envisioned as vastly

different from my brontosaur. *Apatosaurus* was not a waterlogged grubber of algae and water lilies, but in fact was a taut, active animal that trod Jurassic floodplains with its neck and extended whiplash tail held high off the ground. *"Brontosaurus"* as I knew the beast—a hulking pile of flesh and bone that bathed in Jurassic swamps—never actually existed. Almost everything about the monstrous creature—its lifestyle, its skull, and, most regrettably, its name—were human inventions drawn from prehistoric skeletons that actually supported a different form. I had been fooled! The dinosaur *I* met was a petrified museum zombie, shuffling on even though scientists had shot it down decades before.

You see, the dinosaur's major makeover wasn't easy, and it wasn't fast. I had encountered the brontosaur only as it was slowly fading from books and museum halls. A few years before I made my first museum visit, a groundswell of scientific interest in sauropods, stegosaurs, tyrannosaurs, and their varied kin—given the dramatic title "The Dinosaur Renaissance"—had crushed the image of dinosaurs as stupid, abominable reptiles and recast them as animals that had more in common with birds than with any lizard or crocodylian (a term for the group encompassing alligators, crocodiles, and gharials). The fossil bones were the same as they ever were, but paleontologists saw the petrified remnants in a new light. And in the special case of *"Brontosaurus,"* the dinosaur's name, skull shape, and cultural identity are all bound together in a complicated knot where science and imagination meet.

The story started over a century ago during one of the most fruitful times in the history of paleontological discovery. In 1877, the Yale paleontologist Othniel Charles Marsh applied the name *Apatosaurus ajax* to the partial skeleton of a juvenile sauropod that Arthur Lakes, later one of Marsh's field assistants, had discovered in Colorado. Two years later, Marsh coined *Brontosaurus excelsus* on the basis of a more complete skeleton his men had found, this time at Como Bluff, Wyoming.

The dinosaurs were only subtly different, but in Marsh's day,

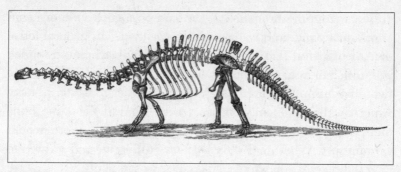

In 1896, the paleontologist O. C. Marsh published this reconstruction of "*Brontosaurus*" *excelsus* in his major monograph *The Dinosaurs of North America*. (Image from Wikimedia Commons: en.wikipedia.org/wiki/File:Brontosaurus_skeleton_1880s.jpg)

paleontologists interpreted even the slightest of skeletal differences as indicators of previously unknown genera and species. After all, Marsh and his contemporaries were among the first to scientifically catalog a prehistoric lost world full of creatures no one had ever seen before. Who could say how many different forms there were?

In this case, though, what Marsh thought were two different dinosaur genera were merged into one. In 1903, the paleontologist Elmer Riggs argued that Marsh's "*Brontosaurus*" wasn't distinct enough from *Apatosaurus* to justify a new genus name. The "*Brontosaurus*," Riggs reasoned, was only a new species of *Apatosaurus*, and since *Apatosaurus* was named first, it had priority of title. Thus "*Brontosaurus*" *excelsus* became *Apatosaurus excelsus*. The trouble was that the name change didn't filter from technical journals to pop culture (or, clearly, to museum displays). As institutions such as the AMNH erected *Apatosaurus* skeletons, they slapped the mounts with the old "*Brontosaurus*" label for reasons that have never been resolved. Maybe they thought the old name sounded better, or were unsure about rebranding one of the most famous dinosaurs in their halls. Whatever the reason, "*Brontosaurus*" was given a second life.

For the moment, let's follow the lead of Riggs's stubborn contemporaries and call the animal *"Brontosaurus."* In general form, the *"Brontosaurus"* skeletons museums so proudly displayed weren't very different from other huge sauropods, such as *Diplodocus*. These two dinosaurs—who lived alongside each other in western North America about 150 million years ago—shared the same body plan, with *"Brontosaurus"* being a bit bulkier than its more slender counterpart. What made *"Brontosaurus"* different, and seemed to characterize the dinosaur's personality, was its skull.

When I met the skeletal *"Brontosaurus"* in 1988, the dinosaur's neck was capped with a skull that made it look about as dumb as early-twentieth-century scientists insisted the animal must have been. As the AMNH paleontologist William Diller Matthew wrote, "We can best regard the *Brontosaurus* as a great, slow-moving animal automaton, a vast storehouse of organized matter directed chiefly or solely by instinct, and to a very limited degree, if at all, by conscious intelligence." To my mind, this man, who oversaw the construction of the mount I was so fascinated by, viewed the dinosaur as a bad evolutionary joke, a heavyweight that was all brawn and no brain.

Unbeknownst to me at the time of my first museum trip, this reptile's skull was a conglomeration of bone fragments and speculation.

When Marsh's field crew discovered the original *"Brontosaurus"* material at Como Bluff, they were frustrated that the specimen lacked a cranium. (Sauropods had a habit of losing their heads between their death and burial.) So, when it came time for Marsh to commission an illustration of what the animal's skeleton would have looked like, he drew on several skull bones found at *another* Como Bluff quarry. These pieces actually came from a different animal—a short-snouted, high-skulled sauropod called *Camarasaurus* that lived at the same time—but Marsh didn't know that.

He assumed that the skull and skeleton belonged to the same animal, and so he used the fragments to re-create a *"Brontosaurus"* skull. Other museums followed suit. It was years before anyone found the dinosaur's true skull.

The beginning of the end for *"Brontosaurus"* goes back to Dinosaur National Monument, one of the richest boneyards ever found. You know you're getting close to the park when goofy, tourist-trap dinosaurs start appearing along Highway 40 in Vernal, Utah. You can't miss them. Some of them snarl, others pose outside hotels, and my favorite—a rendition of the town's long-necked mascot Dinah—wears a polka-dot bikini and stands above a sign that reads: "Let's swim!" Dinosaurs didn't have mammary glands, so I'm not sure what good a bikini top would do. Maybe that's just the Utah sense of modesty at work.

Vernal's dinosaurs have a gleefully outdated feel. They're mostly holdouts from an earlier era, from a tourism boom after a glass-walled museum was erected over Dinosaur National Monument's quarry of bones in 1957. The protected excavation was the dream of Earl Douglass—the man who struck a rich vein of fossils among the area's rocky hills in 1909, and extensively quarried the site under the employ of Pittsburgh's Carnegie Museum of Natural History. Even though Douglass shipped tons of bones back east, he wanted the enormous bonebed to become a living museum where visitors could come and see paleontology in action. Some of his dreams—such as an airstrip on the site and fine dining for high-class clientele—didn't come to pass, but the heart of his vision was realized and continues to show visitors that prehistory can seem as alien as another world.

Cruise past decaying rock shops, a few more cracked and faded dinosaurs, and emerald swaths of farmland that spring from the banks of the Green River, and you'll finally arrive at the park. A dopey *Diplodocus* grins sheepishly at visitors from a parking lot just outside the park limits. And if you know your geology, the winding drive to the recently renovated museum is a literal trip through time. Millions of years of deposition, uplift, and erosion cracked the depths of the earth into a series of sharp slices, each sliver older than the last. Remnants of ancient oceans transition into traces of fern-covered floodplains, divided from the vestiges of dune-filled deserts by the incursion of another vanished sea, and so on down through time. Even if you're not well versed in the paleontological particulars, you can follow the changes by color. Each formation is set off from the others by its own range of hues, from mint green to rust red. I could never dream of a more wondrous landscape. This is one of the most beautiful places on Earth.

The road leading up to the quarry wall is a mixture of maroon slices interspersed through grayish-purple stacks. This is the classic color scheme of the Morrison Formation, the roughly 150-million-year-old deposits that herald the presence of dinosaur giants. This was the era of *Stegosaurus, Allosaurus, Diplodocus, Brachiosaurus, Ceratosaurus,* and many other favorites, including—of course—the dinosaur formerly known as *"Brontosaurus."*

Picking out each species on the sheltered quarry wall isn't easy for anyone who doesn't have an encyclopedic knowledge of dinosaur anatomy. The exposed rock face is a logjam of bones created by an unfortunate twist of Mesozoic fate. Dozens of dinosaurs died in a Jurassic drought, and when the rainy season finally broke the dry spell, the bodies of the poor dinosaurs were washed together into this one place. Disembodied limbs and segments of tail are interspersed with isolated skeletal elements in a slurry of tan-shaded bone. Bad luck for the dinosaurs, but a bonanza for Earl Douglass and other paleontologists that followed him.

The quarry was bigger a century ago. The slanted bonebed extended another one hundred feet upward, and another hundred feet to either side. Those portions were uncovered, excavated, and shipped to museums long ago. And while most dinosaurs in the quarry wound up as isolated bones and body parts, Douglass also exhumed a few complete skeletons. In September of 1909, not long after he stumbled on the string of dinosaur vertebrae that first called his attention to this place, Douglass excitedly picked away at what appeared to be a complete *"Brontosaurus"* skeleton. "We evidently have the most complete of the huge Dinosaurs that ever was found, at least I haven't heard of any other so complete as this appears to be," Douglass wrote back to the Carnegie staff. There was even the possibility that the dinosaur's long-lost skull might be at the end of the neck. "I am not sure," Douglass confided, "but believe now that we will get the head."

The arc of the dead dinosaur's body pointed the way for Douglass. After two more months of excavating the skeleton, he found that the dinosaur's neck was thrown backward over the rest of the spinal column—the classic dinosaur death pose. If the skull was there, surely it would have been at the end of the arched neck. Douglass and his crew carefully uncovered the remainder of the neck "with beating hearts," and as he recounted to his boss William Holland, "I could almost see the skull I was so sure of it for was there not a series of 8 cervicals undisturbed and in natural position." But the neck stopped at the third or fourth neck vertebra. There was nothing else. "How disappointing and sickening," Douglass sighed.

Douglass pushed forward regardless, continuing his work through the coming seasons. He even set up a permanent residence among the colorful outcrops, toiling through the brutal summer heat and enduring the winter chills that annually closed his field operations. And while no skull was ever found attached to a *"Brontosaurus"* neck, Douglass turned up a few isolated skulls of the hefty, long-necked sauropod dinosaurs. Most of these resembled the profile of *Diplodocus*. Instead of having blunt heads with spoon-shaped teeth, like *Camarasaurus*, *Diplodocus* had elongated, shallow skulls tipped with a squared muzzle of pencil-shaped teeth.

But Douglass wasn't entirely sure all the skulls really belonged to *Diplodocus*. Perhaps some of the craniums in his collection truly belonged to "*Brontosaurus*"—still a headless dinosaur at the time. "Can we be positively certain that the supposed skull of Diplodocus is not that of Brontosaurus?" he wondered. Specifically, in 1910 Douglass discovered a puzzling skull very close to the neck of a second "*Brontosaurus*" specimen simply titled No. 40. Douglass believed that the fossilized cranium was a *Diplodocus* head that had rolled away from its owner after death, "though I would be glad to give a fellow [a] crown of glory if he'd convince me otherwise." He wasn't quite ready to go out on a limb and say that he had found—at long last!—the head of "*Brontosaurus*."

William Holland didn't think Douglass had found another *Diplodocus* noggin. He believed that his man in the Jurassic had indeed found the long-lost "*Brontosaurus*" head. The skull was similar to that of *Diplodocus*, so it was no surprise that Douglass's doubts threw him off the trail, but the "*Brontosaurus*" cranium looked a little wider and bulkier, befitting the heavier stature of the dinosaur. Holland argued that the established "*Brontosaurus*" skull form—assembled by Marsh from only a few fossil scraps— was totally wrong. The stout skull Douglass had found near dinosaur No. 40, Holland thought, truly did belong to the "deceptive lizard," properly called *Apatosaurus*.

Scientific uncertainties and paleontological politics continued to complicate the legacy of *Apatosaurus*. Despite arguing that Douglass had finally found the dinosaur's skull, Holland decided to leave his museum's mount of the dinosaur headless, and the reconstruction stood that way for twenty years. Only in 1934, two years after Holland passed away, did the Carnegie *Apatosaurus* get a head, and it was a blunt *Camarasaurus*-like stand-in. No one seems to know who made the decision, but the choice reflected the consensus of the time—similar to Marsh's view—that *Apatosaurus* was a close relative of *Camarasaurus*. Due to their presumed affinity, the two dinosaurs would be expected to have similar skulls. The Carnegie dinosaur, along with its counterparts at Yale and the AMNH, smiled at visiting masses with a substitute head for years. And that peculiar skull Douglass had found next to skeleton No. 40 was placed in the collections of the Carnegie with *Diplodocus* on its label.

Eventually, Douglass's tentative hunch and Holland's assertions were proven correct. In 1975, a physicist turned self-trained sauropod expert by the name of John McIntosh reviewed the various letters, notes, and quarry maps Douglass had composed and confirmed that Douglass's strange *"Diplodocus"* head was found right up against an *Apatosaurus* skeleton. McIntosh described the skull in a paper outlining the fossil's distinct anatomy, and finally put the last, essential piece of *Apatosaurus* in place. On October 20, 1979, the Carnegie officially replaced the incorrect skull with a cast of the rediscovered *Apatosaurus* fossil. Other museums took a little bit longer to fix their mounts. Yale's Peabody Museum exchanged skulls in 1981 ("This is the first head transplant that I've ever performed," the paleontologist John Ostrom quipped as he put the new skull on the skeleton), and the AMNH fixed theirs much later on, during the mid-nineties revamp.

Sure, paleontologists already knew that *Apatosaurus* was the proper name for the dinosaur when the skull switches happened. Riggs had settled that issue in 1903, and various papers cemented the technicality, but even though Riggs made it crystal clear, *"Brontosaurus"* lived on. *Tyrannosaurus rex* might be the undisputed favorite now, but *"Brontosaurus"* ruled the early days of cinema and has left a sizable imprint on the cultural landscape. *Gertie the*

Apatosaurus excelsus as we know the dinosaur today. (Illustration by Scott Hartman)

Dinosaur, one of the first animated features, starred a frisky dinosaur based on the American Museum of Natural History's *"Brontosaurus."* More monstrous brontosaurs would later threaten humans in 1925's *The Lost World* and the 1933 classic *King Kong*. (Not to mention Cary Grant's frustrated search for the dinosaur's "intercostal clavicle"—a bone that doesn't actually exist—in 1938's *Bringing Up Baby*.) And that galumphing, sometimes aggressive personality was encapsulated by that fabricated skull. When the proper skull was placed on the dinosaur's body—just as paleontologists were revising the essence of what dinosaurs were—the animal's entire demeanor changed.

By now, we know that *Apatosaurus* is the dinosaur's proper name. If you note the wrong term in front of a young fossil fan, you'll get a swift correction. But you can't keep a brontosaur down. Everyone knows the dinosaur's name and we want *"Brontosaurus"* to exist. Even though some of my paleontologist friends have tried to match the name's popularity by spreading the name of a previously unknown sauropod, *Brontomerus*—or "thunder thighs"—there isn't going to be another dinosaur that can fill the cultural gap *"Brontosaurus"* left behind, which is funny, since it's not like there's some *"Brontosaurus"*-shaped hole in prehistory. Just look at Google's Ngram Viewer—a service that tracks word use in books through time. We started using *"Apatosaurus"* and *"Brontosaurus"* at about the same time, but the Ngram reveals that *"Brontosaurus"* has always been the victor. Even from the 1970s on, when we *knew* that the dinosaur wasn't real, the name still beats *Apatosaurus* in frequency. Whenever we mention *Apatosaurus*, we feel compelled to remind everyone that the dinosaur used to be called *"Brontosaurus,"* and so the discarded name persists. (I'm certainly compounding the problem here.) We can't conjure *Apatosaurus* without the memory of *"Brontosaurus"* trailing close behind.

The torturous episode reminds me of when Pluto was demoted from planet status to the dwarf planet level. The cosmic body is still out there—scientists didn't destroy it with a Death Star or other interplanetary weapon—but the outcry over the change was intense. Even many die-hard science fans loathed the technical decision. Why should a mundane label change matter so much? As the astronomer Mike Brown, whose work contributed to Pluto's fall from interstellar grace, put it:

> In the days that followed [Pluto's demotion], I would hear from many people who were sad about Pluto. And I understood. Pluto was part of their mental landscape, the one they had constructed to organize their thinking about the solar system and their own place within it. Pluto seemed like the edge of existence. Ripping Pluto out of that landscape caused what felt like an inconceivably empty hole.

The Jurassic herbivore was a touchstone that put the rest of the archosaurian horde in context and helped us revive lost worlds in our imaginations. And the sauropod's apparition remains a cultural baseline against the ever-shifting image of what dinosaurs are. To my mind, we didn't lose a dinosaur so much as gain a much clearer view of a real Jurassic giant. The contrast between old *"Brontosaurus"* and dinosaurs as we know them now shows us just how much we have learned about dinosaur biology.

In order to appreciate how our understanding of dinosaurs has changed, though, we need to know what dinosaurs really *are*. That's not as simple as it sounds. Here's what dinosaurs are not: they are not just anything big, toothy, and prehistoric. A woolly mammoth wasn't a dinosaur, the leathery-winged flying reptiles called pterosaurs weren't dinosaurs, and fish-chasing aquatic reptiles such as the plesiosaurs and ichthyosaurs weren't dinosaurs. Just because an animal's name ends in "saur" doesn't necessarily mean it's a dinosaur. "Dinosaur" is a scientific term, not a colloquial one, and applies only to a restricted group of animals.

The simplest way to visualize this is by picking two of the last members of each branch of the dinosaur family tree and tying them back to their last common ancestor. So if you were to take *Triceratops* and a pigeon (birds are dinosaurs, too) and go back to their last common ancestor, everything that rests within the resulting evolutionary tree would count as a dinosaur, all of them bound together by a mosaic of shared anatomical features. If an animal doesn't fall within those brackets, it's not a dinosaur. That's a strange way to think of delimiting dinosaurian identity, but the proof is in their evolutionary relationships.

Let's dig a little deeper. The reason we pick *Triceratops* and a pigeon to outline the dinosaur family tree is because these animals represent the ultimate members of the two major dinosaur subgroups. The dyspeptic Victorian anatomist Harry Govier Seeley delineated these varieties in 1887 on the basis of dinosaur hips, of all things. While some dinosaurs (such as *Allosaurus* and *Apatosaurus*) had roughly lizard-shaped hips, others (such as *Stegosaurus*) had what Seeley thought were bird-like hips. He named the two varieties the Saurischia and Ornithischia, respectively (even though the latter name turned out to be ironic—although birds are dinosaurs, so-called bird-hipped ornithischian dinosaurs weren't anywhere close to avian ancestry).

While the names don't exactly roll off the tongue, Ornithischia and Saurischia are essential labels for understanding who's who among the dinosaurs. All the dinosaurs we know of fall into one group or the other. The myriad of bizarre dinosaur forms is staggering. Among the Ornithischians were dome-heads like *Pachycephalosaurus*; shovel-beaked hadrosaurs such as the crested form *Parasaurolophus*; armored dinosaurs such as *Ankylosaurus*; and *Pentaceratops*—a massive quadruped with curved brow horns and a flashy, elongated frill. As far as we know, all of these dinosaurs were principally herbivorous.

The Saurischia, on the other hand, includes some of the largest, fiercest, and most charismatic dinosaurs of all. The two principal saurischian subgroups were the sauropodomorphs— long-necked herbivores that included *Apatosaurus* and its close kin—and the theropods. For a long time, "theropod" was synonymous with "carnivorous dinosaur," but that isn't true anymore. *Tyrannosaurus, Allosaurus,* and *Giganotosaurus* were all flesh-rending theropods, as were *Velociraptor* and its kin, but many theropod lineages became either omnivores or herbivores, and those include birds. While the carnivores have traditionally stolen the show, the weirdest theropods belong to recently discovered groups such as the alvarezsaurs—turkey-size dinosaurs thought to be the Mesozoic equivalent of anteaters—and potbellied feathery herbivorous dinosaurs with insanely long hand claws, called therizinosaurs.

Our understanding of just how wildly divergent dinosaur body plans were is constantly changing. The word "dinosaur" technically includes everything from an Emperor penguin to one-hundred-foot behemoths such as *Supersaurus,* heavy-skulled bonecrushers like *Tyrannosaurus,* and spiky, armor-plated enigmas such as *Stegosaurus.* We probably don't even know the full span of dinosaur body types. Within the past three decades alone, paleontologists have identified several kinds of dinosaurs that we had no conception of before. The ant-eating alavarezsaurs and totally weird therizinosaurs are two such groups, but there are also the abelisaurids—theropods with short, deep skulls and wimpy arms that even a tyrannosaur would laugh at—and croc-snouted, sail-backed carnivores called spinosaurs.

And that's to say nothing of the dinosaurs that lived after the mass extinction that closed off the Cretaceous, about 66 million years ago. Dinosaurs were not exclusively prehistoric animals—we now know that birds are the sole surviving dinosaur lineage. Indeed, birds are dinosaurs, but the majority of forms—the types that most immediately spring to mind when you think of the word "dinosaur"—are called non-avian dinosaurs. Many writers and paleontologists prefer to consider "non-avian dinosaur" and plain old "dinosaur" as synonyms because of the cumbersome jargon, but I think it's about time we came to terms with the technical language. Yes, it can be a little unwieldy, but we insult dinosaurs if we ignore the fact that they are still with us.

To most people, "dinosaur" is something extinct. And recent

discoveries—such as the spinosaurs and alvarezsaurs—are showing us how much there is left to be uncovered. Many of these discoveries have come from sites in South America, Africa, and Asia
that were beyond the reach of early fossil hunters, but even North
America and Europe—the continents that have been systemati-
cally sampled for the longest time—have yielded strange dinosaurs unlike anything anyone has seen before.

All these fossil finds come from a distinct swath of prehistoric
time. The Mesozoic span of the dinosaurs ran for more than 160
million years the world over. The dinosaurian heyday fell across
three different geological periods—the Triassic (250 to 200 million years ago), the Jurassic (199 to 145 million years ago), and the
Cretaceous (144 to 66 million years ago). That is a lot of time for
evolution to usher new forms into existence. Even though we may
never find all the dinosaur species, as some probably lived in
habitats where there wasn't the right combination of factors for
fossilization, there are certainly thousands of as-yet-unknown dinosaurs waiting to be found.

Dinosaurs aren't *only* prehistoric animals, real monsters, or
even objects of scientific scrutiny. They're icons and cultural celebrities. As the journalist John Noble Wilford wrote in *The Riddle
of the Dinosaur,* "Dinosaurs, more than other fossils, are public
property, creatures as much of the public imagination as of scientific resurrection." Dinosaurs invade our music, our movies, our
advertising, and our idioms (although "going the way of the dinosaur" should really mean becoming undeniably awesome, rather
than sinking into inevitable extinction). NASA even shot dinosaurs into space twice. Don't ask me what for, but they trans-
ported dinosaur fossils into space all the same—maybe because
the creatures have so utterly entranced us and there's hardly a
higher honor for our favorite monsters than for their bones to be
granted a cherished place on a trip outside our atmosphere.

With dinosaurs everywhere, it's no surprise that going through
a "dinosaur phase" is a common and almost expected part of

American culture. There's something about these creatures that has an immediate and inextricable appeal to children, and more than a few young dinosaur fans hold on to that passion to become paleontologists. I've never heard a good explanation for why this is. I don't buy the pop-psychology logic that dinosaurs are so celebrated because they are animals that are big and fierce, but safe because they're extinct. The appeal of dinosaurs doesn't just lie in our ability to conjure them up and banish them at will. There's something else at work, embedded in our curiosity about where we fit in the history of the world.

Indeed, dinosaurs fueled rampant speculation about history and our place in it even before they had a name. From the Greeks to Native Americans, ancient cultures and aboriginal people concocted legends of hoary terrors and powerful heroes to explain the unusual animal bones they found crumbling out of the earth's crust, and the first English naturalists to describe dinosaurs saw them as fearsome, sharp-toothed reptiles of untold destructive power. Their remains were so strange and frightening that we instantly recognized they were primordial beasts that vanished long ago. More than anything else, the attractive essence of the dinosaurs lies in their bizarre and terrifying nature. We can't help wonder about creatures that, from the very start, we've envisioned as Tennyson's "Dragons of the prime, / That tare each other in their slime."

Those images of dinosaurs easily become entrenched in our minds, even as science continues to revise what we thought we knew about them.

Our understanding starts with finding the dinosaurs themselves. We can't begin to reconstruct the identity of dinosaurs, and the details of their lives, without first collecting their bones.

I thought about this undeniable fact, and the romance of fossil discovery, in 2011 as I stood on the balcony overlooking Douglass's

A view of life in Jurassic Utah, 150 million years ago. This was the habitat of *Apatosaurus*—the bulky sauropod seen crossing the floodplain at center. (Art by Robert Walters and Tess Kissinger, courtesy of Dinosaur National Monument, Utah)

old quarry, now cleaned and dusted to show off the mass graveyard in detailed relief. This is the heart of Dinosaur National Monument, and the crowded graveyard exemplifies the very beginning of our struggle to reconstruct dinosaur biology. The vista is the result of countless hours of work. For years, experts picked away at the rock face to expose the bones right in front of museum visitors.

Today, the work has stopped. Almost everything there is to find has been uncovered, and I'm a little disappointed that I can't watch the diligent fossil excavators go about their work (or even have a crack at carefully chipping out a few bones myself). Finding and digging dinosaurs is grueling, sweat-drenching work, punctuated by brief periods of excitement. When I'm in a quarry, in the lab, or in the field looking for dinosaurs in the rough, uncovering a fossil is an exhilarating experience—when my eyes settle on a freshly exposed bone or fragment, I can't help but wonder what sort of animal it belonged to and where it fit in the organism's skeleton. As George Gaylord Simpson, one of the greatest paleontologists of the twentieth century, once wrote:

> Fossil hunting is far the most fascinating of all sports. I speak for myself, although I do not see how any true sportsman could fail to agree with me if he had tried bone digging. It has some danger, enough to give it zest . . . and the danger

is wholly to the hunter. It has uncertainty and excitement and all the thrills of gambling with none of its vicious features. The hunter never knows what his bag may be, perhaps nothing, perhaps a creature never before seen by human eyes. Over the next hill may lie a great discovery! . . . The fossil hunter does not kill; he resurrects. And the result of his sport is to add to the sum of human pleasure and to the treasures of human knowledge.

That same spirit is what led Earl Douglass to devote his life to uncovering his great Jurassic bonebed, and this romance fueled the institutional "My dinosaur is bigger than yours" contest that yielded splendid reconstructions of *Apatosaurus*—née "*Brontosaurus*"—in the exquisite museum halls of Pittsburgh, Chicago, and New York City. Those fantastic displays, like those all over the world, were petrified trophy rooms that showed off what toiling in the badlands might teach us about prehistory and our place in nature. The delicately mounted skeletons speak of a past so far beyond the reach of human memory that we can't even fully comprehend the depth of that time, and their stock-still skeletons place our own existence in context. (Consider this: *Tyrannosaurus* lived closer to us in time—66 million years ago—than it did to *Apatosaurus*, which lived 84 million years prior.) Indeed, although Douglass's dinosaurs gave his quarry its fame and eventual protection as a museum, the teeth and bones of tiny mammals are needles in this dinosaurian haystack. Our ancestors and cousins snuffled through the undergrowth and hid in the darkness of the Jurassic world, without so much as an inkling that the seemingly indomitable reign of the dinosaurs would someday end.

As fun as fieldwork can be, though, paleontology is far more than a trophy hunt. Finding a dinosaur is only the very start, and the little secret of fossil hunters is that if you pick the right geological setting, and can tell bone from rock, it's not all that hard to find dinosaurs. The exercise relies almost as much on luck as on science. And, after finding a few myself, I discovered that uncovering a dinosaur bone doesn't feel quite the way it's so often portrayed on television. In the endless stream of dinosaur documentaries I watched growing up, a paleontologist would often rhapsodize about being the first pair of eyes to see the freshly uncovered dinosaur bones in 66 million years or more. The scientists were apparently enthused just by the success of the hunt.

But that's not what I thought about as I carefully scraped the sediment off a dinosaur femur in Ghost Ranch, New Mexico; when I plucked up a handful of dinosaur teeth in the ranchlands outside of Ekalaka, Montana; or even when I stared at Dinosaur National Monument's beautiful Jurassic cemetery. Dinosaur fossils are vestiges of ancient life. So many questions can be asked of even a single bone: how the dinosaur moved, what colors adorned its skin (or feathers), what the creature ate, how it died, and where it fit in the wider panorama of life on Earth, just to start. That's the passion of paleontology. Dinosaurs are old, yes, but they are also mind-bogglingly strange. The persistent questions about how such creatures could have evolved and thrived for so long are what drive me, and many other dinosaur fans, to keep digging into their history. And this is not just a dry academic exercise. This is personal. If I can uncover the secrets of their evolutionary success, maybe I can start to comprehend my endless fascination with them.

I had so many questions as a child that I was told we'd never answer. Slowly, and amazingly, we're starting to envision dinosaurs as they truly were. Paleontologists are painting a more intimate portrait of dinosaurs than has ever been composed before. The days of headhunting for prize skeletons and then leaving those bones to collect dust on museum shelves are over. The bones now form the basis of intense research programs that probe, scan, and dissect the fossil remains for whatever clues we can find about the lifestyles of the fierce and extinct. The Dinosaur Renaissance drastically changed dinosaur imagery, but, as the University of Maryland paleontologist Thomas Holtz, Jr., once told me, it's the new Dinosaur Enlightenment that is outlining the details of how the animals lived.

Science is not just the stepwise accrual of facts that are written down and then forgotten. Fact and theory are intertwined, fostering our ever-changing perception of nature. The more we learn about dinosaurs, the stranger they become and the more questions we have about their biology. And the mystery of the dinosaur is wrapped up in two complementary themes—how they lived and why almost all of them disappeared. In order to solve those conundrums, we need to solve a slew of other dinosaur mysteries— how they mated, grew up, and communicated with each other through sound, smell, and sight.

Of all these puzzles, how dinosaurs came to rule the world is one of the most enduring. Douglass's quarry preserves the heyday of giant dinosaurs—a time when a fantastic array of huge herbivores arched their elegant necks over fern-covered floodplains and tried to avoid an almost equally diverse complement of giant knife-toothed predators. For me, at least, this slice of Jurassic time is the acme of the dinosaur's reign. This is a Jurassic classic. Marvelous as it is, though, the tableau contains a thread we can trace back to the mystery at its source. How did *Apatosaurus* and its varied kin get their start? How, exactly, did dinosaurs rise to rule the world in such flamboyant fashion? To find out, I have to look elsewhere, and the first stop is a few rock exposures over in the same park.

A few miles down Dinosaur National Monument's main drag is a little turnoff for the Sound of Silence Trail—a hike that always puts the Simon and Garfunkel song in a near-endless loop in my head as I walk past the low scrub and sandstone exposures beautifully carved by wind and water. Here, at a little kink in the trail, a deep swath of rust-red rock juts out of the ground in a long curve locally called "the Racetrack." In this section of 220-million-year-old time, among preserved ripple marks and tunnels left behind by ancient worms, are the tracks of svelte, gracile dinosaurs—early members of a dynasty that had not yet come to power. The traces dinosaurs left along muddy lakeshores are all that is here, and even those telltale footprints are rare signs of creatures that were only a marginal part of the prehistoric eco-system. To understand dinosaur lives, we must examine these Triassic rocks—a time tens of millions of years before *Apatosaurus* and other dinosaurs stomped and bellowed their superiority. If we are truly going to appreciate dinosaurs for what they are, we have to go back to their humble beginnings.

Two

The Secret of Dinosaur Success

The worst dinosaurs I have ever seen stand alongside Arizona's stretch of I-40. The withered horrors bake in the sun outside Stewart's Petrified Wood Shop, not far from the turnoff for Petrified Forest National Park. They are the cartoonish essence of dinosaurs I encountered as a child—all green skin and horrible teeth. One of the dreadful dinosaurs—what I can only assume was a shoddy attempt at a tyrannosaur—cradles in its jaws a frayed female mannequin with a red shock wig, and another weather-beaten model sits on a decaying sauropod, strapped in by a tangle of icicle lights. It's Barbie meets *The Beast from 20,000 Fathoms*.

Stewart's isn't the only highway stop to use dinosaurs as bait. A few more pit-stop misfits stand alongside the highway on the route to the nearby national park. If you want to get a driver's attention, put out a dinosaur. Grotesque as they are, the sculptures make me wonder what really defines this famous group of animals. The misshapen roadside statues are clearly supposed to be dinosaurs, but many of them don't look at all like the actual creatures paleontologists reconstruct. I try to think of some feature that unites all the sculptures, a common dinosaur denominator that connects all the terrible highway-side forms. I can't pick anything out. They all have a cast of dinosaurishness about them, but why? The question sticks in my mind, but I don't pull off the highway to give the dinosaur wannabes further consideration.

On this clear October morning, I have a date to meet with the Petrified Forest paleontologist Bill Parker.

I was on my way home from a science writing conference in Arizona, but couldn't resist the opportunity to take the wide detour, chat with Parker in person, and take a peek at the dinosaurs hiding in the Petrified Forest's collections. It would be a great chance to learn more about the creatures that inspired the awful highway Americana and to piece together two intertwined mysteries—what a dinosaur actually *is*, and what allowed dinosaurs to thrive for so long.

I had made Bill's virtual acquaintance a few years before. In the tiny circle of paleontology blogs, Bill's is a great one. He runs Chinleana, which focuses on his own research and related studies. The blog's name comes from the Chinle Formation, an expanse of 228- to 200-million-year-old Late Triassic time that's exposed at Petrified Forest and elsewhere.

By the time I roll up to the park's gift shop and visitor center, Bill has just a few minutes to spare. An administrative meeting sometimes takes precedence over a dinosaur fanatic. But that's okay. Even five minutes among the park's collection of fossils is worth the trip. On our way past the park's offices to the fossil preparation and storage space, I explain to Bill that I've stopped by to learn more about Petrified Forest's early dinosaurs. Bill knits his eyebrows. "Well, we don't have that many," he says. He leads me to the park's fossil collection to explain why.

Beneath the storeroom's blue-white fluorescent lights, Bill kneels down and unlocks a squat olive-green safe—the holotype cabinet. This is the secure box where the park keeps important fossil specimens that have been used to establish new species of Triassic life. Bill slides out a drawer containing what seem to be mundane scraps of bone, but those small pieces are the chief representatives of an enigmatic early dinosaur named *Chindesaurus*. The fossil is distinctive enough to deserve its own name, but so little is known of the animal that it's difficult to tell exactly what it would have looked like or what its closest relatives were. Scattered fragments, like those resting in the cabinet drawer, are all that paleontologists have found so far. And that is actually typical of most dinosaurs. Contrary to what Steven Spielberg's adaptation of *Jurassic Park* depicted, paleontologists usually find only isolated bones and fragments of dinosaurs. Partial skeletons are rare, and complete, articulated dinosaurs are even more elusive.

That's why paleontology relies on comparative anatomy to reconstruct ancient life. Scientists compare scrappy skeletons to more complete ones from closely related forms to fill out the overall appearance of the whole animal. The trouble is, sometimes there isn't a particularly close relative to serve as a stand-in, and that's the case with poor incomplete *Chindesaurus*. We know a bit more about the only other dinosaur found in Petrified Forest. Called *Coelophysis*, this was a slender carnivore that had much to fear from bigger, more powerful predatory neighbors. The dinosaur was not anywhere near as big or as gruesome as the terrible models that stand a few miles outside the park.

Until recently, paleontologists thought that there might have been a third dinosaur in Petrified Forest. After locking up the holotype drawers, Bill points out a couple of skull casts sitting on the top of a long row of cabinets running down the room. The reconstructed heads could fit comfortably in the palm of my hand, and they look nothing like that of any animal alive today. The restored cranium looks a bit like a deep alligator skull that has been smushed into a shorter form, with rounded teeth. This is *Revueltosaurus*, Bill says, the "dinosaur" that wasn't.

For years, paleontologists assumed that *Revueltosaurus* was an elusive form of dinosaur. Adrian Hunt named the animal in 1989, when *Revueltosaurus* was represented by nothing more than a collection of battered teeth. Based on their structure, however, Hunt thought that they were once set in the mouth of a particular form of early ornithischian dinosaur (a precursor of the horned ceratopsians, shovel-mouthed hadrosaurs, and other forms that would come later). This was important. While paleontologists had found definitive skeletons of saurischian dinosaurs in the Triassic of North America (such as *Coelophysis*, an early relative of carnivores such as *Allosaurus* and a distant relative of the titanic sauropods), no one had ever found an ornithischian. Up until this point, half the dinosaur family tree appeared to be missing from the continent during Triassic time. All Hunt had was teeth, but those teeth hinted to him that there were skeletons just waiting to be found. But when a *skull* with teeth was finally found, it turned out that *Revueltosaurus* wasn't a dinosaur at all. The teeth fit into the mouth of a very different animal that was more closely related to crocodiles than dinosaurs.

Just as ornithischians and saurischians were kinds of dinosaurs, the entirety of Dinosauria was just one branch in a wider family tree called the Archosauria—the justly named "ruling reptiles." Today there are only two types of archosaur left—birds and crocodylians. During the past 250 million years, though, there were many, many other types of archosaurs, all of which died out at one time or another.

Birds and crocodiles are our modern guides to this family tree. The Archosauria is typically split into those forms that were closer to birds (the Avemetatarsalia) and those that were closer to crocodiles (the Pseudosuchia). The crocodile side is the one that *Revueltosaurus* is anchored to, and Parker, with a few colleagues, concluded that it was probably a primitive relative of the aetosaurs. These well-armored archosaurs looked like a crocodylian version of a pig, and a pig decorated with bone plates and spikes at that. And these long-lost croc cousins were just one marvelous form of life peculiar to the Triassic.

Chinle time was an age of wonderfully fantastic life. If you could travel back to this part of the Triassic you'd see many vaguely familiar creatures, harbingers of what was to come, but they wouldn't look exactly like anything we see around us today. The dinosaurian ancestors of birds were rare, slender creatures; the closest relatives of the first mammals were either hulking tusked holdovers from an earlier era or tiny shrew-like fuzzballs; and a profusion of bizarre crocodile relatives ruled the land, from armor-encased omnivores to aquatic ambush predators and bipedal dinosaur mimics. The Late Triassic was when reptiles began to rule, but the earliest dinosaurs only hinted at the potential of what was to come. Growing up, I had often heard that the Triassic was the "Dawn of the Dinosaurs," but the more I've learned, the less apt this title seems. Triassic dinosaurs were the prehistoric equivalent of Chekhov's gun—loaded with potential, but not set to go off until much later in the world's evolutionary story.

My visit with Bill just makes me anxious to see more fossils, especially the bones of the archosaurs that reigned before dinosaurs dominated the planet. I had come here looking for dinosaurs, but seeing the skull of *Revueltosaurus* reminded me that there was far more to the Triassic story than my favorite prehistoric animals.

Fortunately, a few of the weird Triassic creatures are featured in another Petrified Forest National Park museum farther down the park's main roadway. I cruise past the desert hills, thinking about what life must have been like during this time. The crumbled remnants of Triassic trees—red and purple fossil chunks against the gray sediment—have tumbled out of the hills and road cuts to create a surreal vestige of prehistoric forests. Here, the forest primeval has gone to pieces. I want to ditch my car at a turnout and wander among the fossiliferous hills, but that will have to wait for another day. I have only an hour to explore the Rainbow Forest Museum before I need to drive back to Utah.

There aren't any dinosaur skeletons among the park's brightly lit displays. The Rainbow Forest Museum makes it abundantly clear that when the fossil-bearing rocks of Petrified Forest were laid down, the croc-line archosaurs ruled the land—as long as you read the signs and don't assume that any extinct creature with big

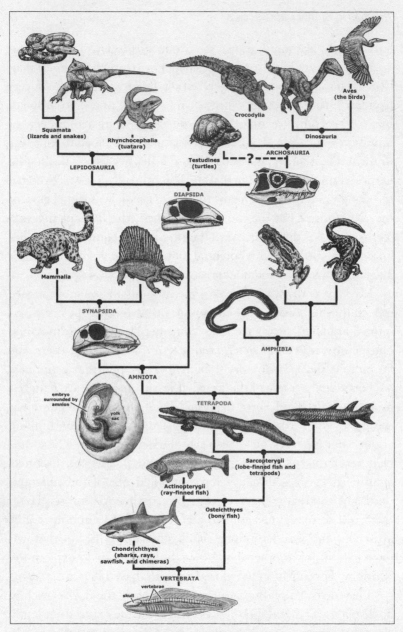

The vertebrate family tree. Archosaurs—the greater group to which dinosaurs belong—is at the upper right. Crocodylians and dinosaurs (including birds) are evolutionary cousins within the Archosauria. (Illustration by Jeffrey Martz)

teeth was a dinosaur. There are a few skeletons that look weird and ancient enough to be dinosaurs, but they actually represent very different animals. Just beyond the small gift shop is a phytosaur skull in a glass case. The creature's head resembles that of a crocodile, but from years spent poring over books and technical papers, I know that this isn't a crocodile at all. The easiest way to tell is by the placement of the nostrils, which are at the tip of the snout in crocodiles, but up near the eyes in this trap-jawed archosaur. And, really, we should call crocodiles "phytosaur-like" rather than the other way around. These aquatic ambush-predators perfected the wait-and-strike aquatic ambush tens of millions of years before crocodiles would do the same.

I notice a trio of skeletons menacing one another in the next display over. There are no dinosaurs here, either. On the left squats *Placerias*—a tusked, beaked quadruped more closely related to us than to any reptile. This dopey-looking animal was one of the last of its kind, a holdover from a time when the forerunners of mammals were the dominant creatures on land. This herbivore, technically called a dicynodont, had a skeleton that looks like an osteological barrel with sturdy limbs and a tusked turtle skull attached. This was the big, plodding plant eater of its era, and it was undoubtedly food for the park's Triassic apex predator, *Postosuchus*. The terror of Petrified Forest was a deep-skulled, knife-toothed carnivore most closely related to the forerunners of crocodiles. *Postosuchus* looked like a crocodile's impression of a tyrannosaur—a bulky animal with legs tucked beneath its body, able to run down prey on two legs as well as four. This is the stuff nightmares are made of.

On the right side of the diorama is *Desmatosuchus*. This was an aetosaur, or, as those who study these creatures lovingly call them, an "armadillodile." The name fits. Covered in long sets of bony plates and bearing a set of curved spikes over the shoulders, this blunt-nosed creature was an omnivore that sliced plants and grubbed for food in the ancient dirt. Not all the prehistoric croc forerunners were out for blood.

These Triassic players, and their neighbors, come to life in the Rainbow Forest Museum's next room. A recently installed mural by artist Victor Leshyk presents the whole Triassic fauna. The painting, simply titled "Late Triassic Life," hardly contains any dinosaurs at all, but embodies an essential lesson about the Triassic. A phytosaur floats in the foreground, an aetosaur snuffles around on a distant bank, a *Postosuchus* snags a hapless little archosaur by the tail, and, right in the middle, another phytosaur bursts out of

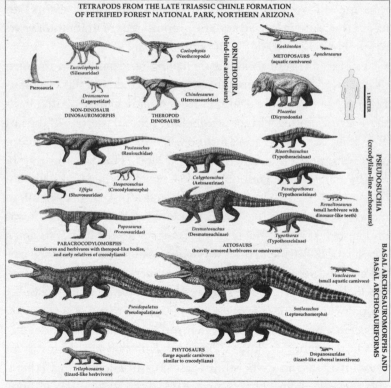

A gallery of Late Triassic creatures from Petrified Forest National Park's Chinle Formation. The snap-jawed phytosaurs (along the bottom), heavily armored aetosaurs (in the middle), and varied other archosaurs (such as *Postosuchus*, *Effigia*, and *Poposaurus* at left) ruled the land. There were only two dinosaurs here—*Coelophysis* and *Chindesaurus* (at top center). (Illustration by Jeffrey Martz)

the water to snatch a *Coelophysis* from the water's edge. That's how marginal the dinosaurian presence is at Petrified Forest. The painting doesn't show ravening hordes of *Coelophysis* killing and out-maneuvering the competition. At Petrified Forest, and elsewhere, dinosaurs were pipsqueaks who sneaked by in a world haunted by larger, more powerful creatures.

I don't like seeing *Coelophysis* as phytosaur chow. At the time of the restored scene, around 225 million years ago, dinosaurs had already been around for about 5 million years. The Age of Dinosaurs had already started—didn't the phytosaurs and other would-be predators have enough sense to know that? I know my internal discomfort is silly. From the very start, dinosaurs were parts of complex ecosystems where some acted as predators and even more were prey for other creatures. As I leave the museum and wind my way back through the painted hills toward the inter-state, I imagine the scattered pieces of fossil wood reassembling themselves into a dense, dark forest, not unlike the West Coast's redwood groves. In my mind's eye, a *Postosuchus* patrols the forest floor, and off in the background—just visible as a glimmer of fuzz and scales—a dinosaur skitters away.

The denizens of the Triassic forest didn't bow to the dinosaurs. *Coelophysis* was like a ghost on the landscape, a dinosaur that didn't embody the classic image of reigning magnificence. Ultimately, though, the dinosaurs not only survived but thrived as the croc branch of the archosaur family tree was decimated. Something changed the course of life. What made dinosaurs so special? The mystery continued to nag at my mind as I headed north for Utah. Maybe there is some clue in what makes a dinosaur a dinosaur, I thought.

Dinosaurian identity has always been in flux. In 1824, before anyone could conceive that the creatures we now recognize as dinosaurs even existed, the British naturalist William Buckland described a paltry collection of strange bones under the name *Megalosaurus* ("great lizard"). "Whilst the vertebral column and extremities much resemble those of quadrupeds," Buckland told his colleagues at the Geological Society of London, "the teeth show the creature to have been oviparous, and to have belonged to the order of Saurians or Lizards." This poorly known creature wasn't a modest, squirming reptile, either. Based on the size of the animal's femur, and with the assumption that the reptile was proportioned like modern lizards, Buckland concluded that *Megalosaurus* "would have equalled in height our largest elephants, and in length fallen but little short of the largest whales." The second such reptile to be described (named *Iguanodon* by Gideon Mantell the following year) was an herbivore of equal size. These animals were strange, enormous versions of crocodiles and iguanas from a time incomprehensibly distant from the modern era.

Then Richard Owen made these bizarre animals his own. In 1842, he published an scholarly survey on British fossil reptiles. Instead of considering *Megalosaurus, Iguanodon,* and the more recently described armored form *Hylaeosaurus* as greatly expanded versions of modern lizards and crocodiles, Owen portrayed them as unique, unprecedented reptiles that represented the apex of scaly vertebrate life. United by a unique mosaic of features—from the nature of the spine to the shape of their teeth—these three animals took their place in a new group that Owen coined just for them. They were the Dinosauria, what Owen described as "fearfully great, a lizard."

Ten years later, Owen received the unique opportunity to bring his vision of these animals directly to the public. In 1852, he was tapped to be the scientific advisor to Benjamin Waterhouse Hawkins, an artist who had been commissioned to create life-size models of dinosaurs and other fossil animals for London's Crystal Palace Exhibition. Owen outlined his vision of dinosaurs and schooled Hawkins in their anatomy, and the artist executed the designs. The dinosaurs, resurrected life-size, looked like reptilian versions of rhinos and elephants—like putting a crocodile body on a hippo chassis, wrapped in scaly skin. (The sculptures—now grossly dated—still greet visitors to the Crystal Palace's home at Sydenham Hill.) Owen's versions of *Megalosaurus* and kin were not even close to dinosaurs as we know them today, but instead were strange reptile-mammal hybrids.

Like those before him, Owen had only scraps to work with. The knowledge of his time and Owen's own theories melded together as he tried to create these outlandish forms. But when paleontologists across the Atlantic began to probe the fossil-bearing rocks of New England, they didn't find anything quite like the strange sculptures Owen and Hawkins had created.

In 1858, just a few years after Owen's dinosaurs debuted, polymath Joseph Leidy described the partial skeleton of a herbivorous dinosaur pulled from a marl pit in southern New Jersey. He named the ancient herbivore *Hadrosaurus*. This wasn't a four-on-the-floor model like Hawkins's creations, but an even more peculiar dinosaur with short, slender arms and long, robust legs. A good deal of the dinosaur's skeleton was missing, but, combined with evidence from trackways and other partial skeletons, *Hadrosaurus* forced naturalists to realize that many dinosaurs were bipedal and more birdlike than they ever expected. The American paleontologists Edward Drinker Cope and Othniel Charles Marsh, as well as their British peer Thomas Henry Huxley, began extolling the extremely active, avian nature of dinosaurs. Dinosaurs in artistic depictions still dragged their tails and smiled with a reptilian grimace, but by the 1870s the basic nature of what dinosaurs were had finally been filled out.

Most paleontologists of the time agreed that what set dinosaurs apart was their posture. Whether they towered on two legs or four, they stood erect with their limbs directly beneath their bodies. And for decades, paleontologists thought that feature was the secret of their success. The paleontologist Alan Charig codified this idea in 1972, when he set about figuring out why dinosaurs stood with their limbs erect while crocodiles and their prehistoric relatives, as far as people knew at the time, had arms and legs that sprawled out to the side. He surveyed the hips and legs of various dinosaurs, crocodiles, and their close kin. All of the archosaurs fell into one of three categories. There were the sprawlers (they had legs thrown out to the sides and they crawled, like lizards); animals with a "semi-improved" posture (with upper leg bones held more vertically and placed at a significant angle to the rest of the body, as in crocodiles); and some—like dinosaurs—with limbs held in a column-like fashion directly beneath their bodies. This not only characterized the different archosaur limb postures, but seemed to represent a three-step evolutionary pathway leading to the unique dinosaur posture. Charig's value-based terminology left no question that dinosaurs were superior to everything that had come before. Dinosaurs were unique in that they had a perfected stance that reduced the amount of energy required for each step, which ultimately made them faster and deadlier than anything else. The most conspicuous feature of dinosaurs gave them a competitive edge in a Triassic world where speed made all the difference.

The contemporary, iconoclastic paleontologist Bob Bakker ran with this idea in an article entitled "The Superiority of Dinosaurs." Dinosaurs were not the mental and physical sluggards they had always been portrayed as. Our mammalian bias colored our vision of the past and caused us to underestimate dinosaurs, Bakker argued. During the Triassic, *our* relatives—including the heavy-bodied *Placerias* as well as small, shrew-like creatures—were sprawlers while early dinosaurs were efficient runners. Thanks to natural selection, dinosaurs were simply better engineered, and their skeletal architecture hinted at other features—such as active metabolism and complex behavior—that made them superior to our own forebears. By the time the first mammals evolved during the Jurassic, dinosaurs had already won the day. Dinosaurs got the jump on mammals by millions of years, and Bakker argued that mammals were runners-up in life's race for so long because they "were competitively inferior to dinosaurs during the day and were forced to seek secure diurnal shelters in trees and burrows to escape the great reptiles."

When paleontologists discovered dinosaurs near the ultimate root of the group's family tree, the idea of dinosaur tactical superiority received a major boost. In 1993, just as *Jurassic Park* stimulated dinomania to a fever pitch, the paleontologists Paul Sereno, Catherine Forster, and their colleagues named *Eoraptor*—the "dawn thief"—which they found in roughly 231-million-year-old rocks in Argentina's Valley of the Moon. This was the earliest and most archaic dinosaur ever found, and it sure looked like a consummate predator. Though it topped out at only about three feet long, *Eoraptor* had grasping, clawed hands and a mouth set with pointed teeth. Its contemporary *Herrerasaurus*, discovered in the same valley years before, was even more fearsome. This ten-foot-long carnivore had a boxy skull full of recurved teeth well suited to slicing flesh. Dinosaurs, in their earliest guise, were fast, flesh-rending bastards that totally dominated that landscape.

Soon after these discoveries, the late paleontologist and Mormon bishop William Sill called the upright limb posture of these early beasts their "secret weapon" in the war between dinosaurs and other forms of Early Triassic life. Or, as I recall the narrator of the sensationalized documentary series *PaleoWorld* explaining, "Dinosaurs didn't invent killing; they perfected it." At the outset, Sill and other paleontologists said, dinosaurs were meat eaters that leaped out of the dawn and tore apart whatever they could catch. Only later, once they had fully subdued the competition, did some lineages take up a more peaceful, plant-munching lifestyle.

The cherished tale of a rapid dinosaur coup lives on at the New Mexico Museum of Natural History and Science. I took the ten-hour drive down to the paleo-centered museum to see the imposing mount of the sauropod *Diplodocus hallorum* defending itself against an assault by *Allosaurus maximus*. (These giants—among the biggest dinosaurs of their kind—were once called *Seismosaurus* and *Saurophaganax*, but turned out to be larger versions of already known genera.) To get to the large-scale battle scene, though, you'll need to take a chronological tour through the Triassic exhibit, and about halfway through—in a dark cul-de-sac framed by low couches—an animated short film lays out the basic principles of Triassic life. Natural selection not only explains the dramatic changes in prehistoric life, the narrator's soothing voice relates, but also makes sense of why some creatures—such as lungfish— are so similar to their prehistoric counterparts. There is no inevitable ladder of progress that all species must climb. Natural selection drives nature's most astonishing transformations, but if there's no impetus for change, it can also maintain some forms for millions of years.

As the narrator affably describes, dinosaurs fall into the category of *transcendent* change. Preceded by platoons of squat, waddling creatures, dinosaurs were a major evolutionary improvement that quickly overtook the world. As a cartoon dinosaur stomps on one of its awkward and outmoded ancestors, the narrator explains that "one thing that makes a dinosaur a dinosaur . . . is the fact that the limbs are upright, directly under the body." Standing up straight gave dinosaurs their edge, the traditional story goes, and so early dinosaurs slashed and bit their way to ultimate dominance.

But previous generations of researchers gave dinosaurs too much credit. As paleontologists have uncovered new evidence and reconsidered the old, they've discovered that posture alone couldn't have been the secret to dinosaur success. In fact, dinosaurs were not the only creatures to walk tall. And this change has undermined the traditional understanding of what a dinosaur truly is. An upright posture was thought to be an easy-to-spot symbol of dinosaur identity as well as an explanation for why the creatures became so successful. Now we know that's no longer true, thanks in part to the work of the early archosaur expert

Sterling Nesbitt. I asked Sterling to help me as I struggled to parse the story of the earliest dinosaurs.

I wanted to know what separated dinosaurs from the various other stripes of archosaurs. Since posture is no longer the key factor, Sterling said, we have to move on to other clues to figure out what makes dinosaurs distinct from all their various archosaurian relatives. If you really want to separate the earliest dinosaurs from creatures that are only superficially dinosaur-like, he said, you have to get down to the anatomical nitty-gritty.

There's no stark characteristic that the untrained dinosaur fan can settle on to tell a dinosaur from a cousin croc-like archosaur with upright limbs like *Postosuchus*. One of the key features that can help us determine what's what, Sterling pointed out, is a large flange on the upper arm bone where some of the chest musculature is attached—it's larger in dinosaurs relative to other archosaurs. There are a few other subtle characteristics as well, but as distinct as later dinosaurs were, the earliest ones lived in a world full of evolutionary copycats. Similar features and natural histories evolved multiple times among early archosaurs.

It's counterintuitive, but new fossil discoveries can complicate our attempt to solve the twin mysteries of dinosaur identity and success. As paleontologists fill out the base of the dinosaur family tree and uncover the closest relatives of early dinosaurs, Sterling said, "most of the morphological gaps between dinosaurs and other archosaurs disappear discovery by discovery." Even though the most famous dinosaurs—*Tyrannosaurus*, *Apatosaurus*, and their ilk—were distinct animals that were clearly different from everything else at their time, the earliest dinosaurs were not very different from their ancestors. In a way, we see dinosaurs as unique only because extinction claimed all the similar, closely related archosaurs. This is a fabulous yet intricate confirmation of evolution's grand pattern, but it makes pinpointing the start of dinosaur history maddeningly difficult. There is no isolated, dramatic feature that determines membership in the Dinosauria.

In fact, dinosaurian uniqueness took a major hit when Sterling found an unexpected dinosaur mimic. The cryptic specimen wasn't a fresh discovery, but a forgotten relic extracted from one of the richest Triassic localities in the world. That site is at Ghost Ranch in northern New Mexico.

When you pull up to Ghost Ranch, the place doesn't exactly look like one of the most important dinosaur sites in North America. A quick turn off a scenic road near Abiquiu, New Mexico, the old haunt of desert artist Georgia O'Keeffe is now a retreat center run by the Presbyterian Church. The locale is strewn with bungalows and campsites. For paleontologists, working Ghost Ranch is about as luxurious as it gets—the bathhouses have laundry machines *and* hot showers. A paleontologist could get spoiled in a place like this.

The first time I visited Ghost Ranch, I spent almost a week picking away at the gray, bone-filled sediment of the Hayden Quarry across the road from the main entrance. But the site that made Ghost Ranch world famous among paleontologists lies deeper within the Triassic hills. The University of Utah paleontologist Randall Irmis led me there, along with the rest of a summer field crew, on a low-key afternoon near the end of our 2011 expedition. In a gully flanked by towering orange rock faces, a flow of sandy sediment cascaded away from a damaged stone wall. Flecks of white plaster left over from excavations completed long ago dotted the loose sand. The fossils are gone now. The site was excavated for all it was worth, and a significant portion still remains in plaster jackets, waiting to be cleaned and studied, but the signs of valiant effort remain. This was a mass graveyard of *Coelophysis.*

The *Coelophysis* bonebed is what made Ghost Ranch famous among paleontologists. In 1947, Edwin Colbert scoured Ghost Ranch for fossils. The expedition was supposed to be a quick stop for Colbert's crew on their way to the Petrified Forest, but George Whitaker, a member of Colbert's team, found a few intriguing dinosaur fragments and encouraged them all to stay. As the group carefully picked away at the earth with awls and small tools, even more bones appeared. They had stumbled upon an immense treasure trove of tangled dinosaur skeletons.

Colbert's team stayed at the site from June through September, just enough time to scratch the surface. The dinosaurs were laid down so thick that collecting *individual* specimens was pointless. Huge chunks of dinosaur-bearing rock were extracted from the quarry (one enormous doughnut-shaped block still sits in the Ghost Ranch paleontology museum, still in the process of preparation). Hundreds of *Coelophysis* specimens were recovered from this one spot. No one knows why the dinosaurs accumulated here in such numbers.

But the abundant Ghost Ranch dinosaurs also fell victim to what is both the persistent curse and the blessing of fossil hunting. Field expeditions often collect more than can be carefully prepared and studied, and such was the case with the Ghost Ranch fossils. Well-preserved and complete dinosaurs were prepped for study and display, but there were plenty of plaster jackets that were shelved for analysis in the future. Some sat in storage at the American Museum of Natural History for so long that no one could recall what was actually inside them.

Then Sterling came along. While working as a graduate student at the AMNH, Sterling riffled through the detailed bonebed maps Colbert had drawn in 1947. Sterling was looking for additional specimens of *Coelophysis* he could carefully prepare in the lab to investigate the dinosaur's anatomy. But something else caught Sterling's eye. The map cited a few blocks labeled "phytosaur," and these aquatic predators were relatively uncommon in the quarry. Sterling tracked down the block and began to pick away at the sediment. Only he didn't find a phytosaur. Peeking out of the rock was the hip and foot of a very unusual archosaur.

The creature looked similar to a poorly known archosaur called "*Chatterjea*" that had been described from the Triassic of Texas years before, but Sterling needed more to be sure what he was looking at. To his frustration, however, the jacket containing the front half of the animal—itself preserving the telltale clues to the archosaur's identity—was nowhere to be found. "Another two months went by," Sterling explained, "and the collections manager . . . told me that there was a *Coelophysis* block that had been found hiding among mammoth skulls in the fossil mammal collections."

He wasted no time finding out what the mystery jacket held. "I raced over there and there it was, a partially prepared front half of the '*Chatterjea*'-like animal," he recalled. "I could even see an upside-down skull." Not only did the skull resolve some questions about the relationships of croc-line archosaurs, but the skeleton of this new animal—ultimately represented by four specimens—was extremely dinosaur-like.

In 2006, Sterling and his advisor Mark Norell named the animal *Effigia okeeffeae*.

The true identity of *Effigia* was given away by the archosaur's ankle. Dinosaur ankles are dominated by a large, triangle-shaped bone—the astragalus—and have a very small accessory ankle bone called the calcaneum. Their ankles look like a simple hinge. But crocodile-like archosaurs have a large ankle bone that locks together into a complex unit where the connection between the ankle and foot has an S-shaped divide. This is the kind of ankle *Effigia* had.

50 cm

Dinosaurs weren't the only archosaurs to walk tall. Creatures more closely related to crocodiles, such as *Shuvosaurus*—seen here—independently evolved an upright, bipedal posture that allowed them to swiftly run around the landscape. (Illustration by Jeffrey Martz)

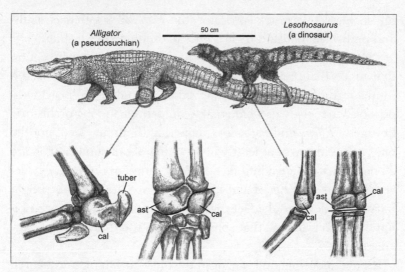

One of the few ways to tell a croc-line archosaur from a dinosaur is by the construction of their ankles. While the pseudosuchians had a complex joint with a large calcaneal tuber, analogous to the human heel, sticking out the back, dinosaurs—such as the ornithischian *Lesothosaurus* here—had a simpler, hinge-like arrangement. (Illustration by Jeffrey Martz)

The articulation with the hip was also key. In dinosaurs, the head of the femur juts inward to a hole in the pelvis. But the femur in *Effigia* articulated with the pelvis in a different way, in a fashion similar to croc-line archosaurs rather than to dinosaurs. The evidence was clear. Though *Effigia* was undoubtedly a crocodile cousin, the animal walked with an upright, bipedal posture like the early dinosaurs it lived alongside. Standing on two legs, this toothless archosaur had a long neck, tiny arms, and a body counterbalanced by a long tail. The deadly and über-efficient skeletal design of dinosaurs was not unique, after all.

Effigia wasn't an evolutionary fluke. The creature was quite similar to another animal from roughly the same time period, named *Shuvosaurus*. And while both *Effigia* and *Shuvosaurus* were toothless bipeds, a lovely skeleton of their close relative *Poposaurus* shows that there were sharp-toothed varieties, too. And all three were offshoots of a line of frightening creatures called rauisuchians—terrestrial predators like *Postosuchus* with deep skulls and limbs held beneath their bodies in dinosaur-like fashion.

So the upright posture of dinosaurs wasn't a unique invention that made them an unstoppable force. "This has always been a funny argument to me," Sterling said. Not only is it "nearly impossible" to recognize evolutionary competition among prehistoric lineages, "*Effigia* and *Poposaurus* show that dinosaurs were not the only game in town, at least when talking about stance." Posture alone wasn't the deciding factor. Why dinosaurs ultimately succeeded, and why *Effigia* and kin didn't leave descendants, might have come down to the fact that dinosaurs "had a unique combination of characters" that somehow gave them an evolutionary advantage.

Some undecipherable factor in the dinosaurian mosaic favored *Coelophysis* and kin over the croc-line archosaurs. By the start of the Jurassic, about 199 million years ago, dinosaurs emerged from the sidelines and took over the ecological roles previously filled by a different cast of creatures. This wasn't the outcome of an unbroken contest between ancestral dinosaurs and other animals in head-to-head combat. Extinction, the often-underappreciated flip side to evolution, may have created a world where dinosaurs could flourish. Even though the non-avian dinosaurs were ultimately undone by extinction, they got their shot thanks to a pair of ecological catastrophes.

Before 250 million years ago, there were no archosaurs. Landscapes the world over were filled with synapsids—the diverse array of creatures more closely related to you and me than to reptiles. Barrel-bodied, tusked dicynodonts grazed in huge herds; gorgonopsians—saber-fanged predators built like excessively muscular dogs—hunted down large prey; and the precursors of the first true mammals—the small, shuffling cynodonts—burrowed and snuffled their way around the Permian world. Then everything went to hell. A mass extinction more devastating than any before or since rapidly winnowed down Earth's biodiversity.

More than 90 percent of the known species in the oceans were lost, as well as more than 70 percent of species on land. By the dawn of the Triassic, complex and thriving ecosystems were replaced by degraded crisis communities populated by just a few species.

Exactly what triggered the Permian extinction is only known in outline. Rapid, intense atmospheric fluctuations and rampant global warming, related to massive amounts of greenhouse gases released into the atmosphere from their geological prisons, are the most likely culprits. Atmospheric oxygen plummeted, the oceans acidified, and the land baked, though not so thoroughly that life was extinguished. There were survivors, and the lineages that *did* make it through were faced with a world both barren and full of possibility. Now that the complicated, interacting ecologies were swept away, surviving lineages were relatively unfettered and could adapt and radiate into new forms. Evolution favored the archosaurs in this new world.

The very first archosaurs originated after the effects of the Permian mass extinction ebbed. While the precise origins of the group are as yet unclear, by 244 million years ago archosaurs were already stalking across the world. Some of the earliest known archosaurs, such as *Arizonasaurus, Xilousuchus,* and *Ctenosauriscus,* looked like crocodilian renditions of a greyhound with prominent sails on their backs. And, according to the growing record of these creatures and revisions of how they are related to each other, it seems that archosaurs diversified very quickly after their origin. While *Arizonasaurus* was strutting through what would one day become the American Southwest, dinosaur forerunners were walking through the area which now houses Poland's Holy Cross Mountains. The bodies of the animals in question have not yet been found, but the AMNH paleontologist Stephen Brusatte and his coauthors have presented evidence from early Triassic footprints that dinosauromorphs were there. The anatomy of the footprints matches the bone structure of the feet of these animals—

not true dinosaurs, but slender long-legged creatures that comprised the group from which the first dinosaurs would later diverge. Dinosaurs, in a definitive sense, wouldn't come onto the scene until later, but the lineage of archosaurs to which they belonged almost instantly split away from other groups in the wake of the Permian catastrophe. Recent discoveries such as *Asilisaurus*—a graceful Middle Triassic dinosauriform with slender legs and an extended neck—at least partly represent the sort of animal from which the first dinosaurs evolved. The defeat of the synapsids at the close of the Permian cleared the way for the progenitors of dinosaurs.

Yet, as I learned during my Triassic road trips, dinosaurs did not immediately take over the world. The first ones were rare, and *Eoraptor* may have been more of a plant-focused omnivore than the terrible little blighter it was originally characterized as. The image change has infiltrated that common purveyor of paleontology products—the CGI-filled basic-cable dinosaur documentary. In the overhyped miniseries *Dinosaur Revolution*, flocks of brightly colored *Eoraptor* are shown as timid little things that have to contend with an especially acrobatic relation of *Postosuchus*, the razor-jawed *Saurosuchus*. Dinosaurs didn't rule the Triassic Valley of the Moon. They eked out an existence alongside more powerful creatures.

During Triassic time, the dinosaur subplot was just one small part of the greater archosaur story. If you traveled back 230 million years, with no knowledge that dinosaurs would eventually reign, you'd probably consider the dinosaurs to be cute creatures of little importance. There was no clue of what was to come. The same could be said of our own kin that lived during the same time. Synapsids didn't entirely disappear. After all, we're the descendants of such proto-mammals that survived into the Triassic.

We don't know why the archosaurs took over the world from our synapsid precursors. Not for certain. But a large-scale analysis of archosaur and synapsid fossils gives us a clue. When researchers tracked the fates of the two lineages after the Permian, they didn't find many signs of direct competition between our primordial relatives and the archosaurs. It was not as if synapsids and archosaurs directly battled each other for control of habitats. Mammals and their forerunners, for the most part, remained relatively small, while the archosaurs diversified into a variety of body sizes— from creatures the size of a pigeon to the largest of the sauropod dinosaurs, reaching one hundred feet in length or more. Biological constraints might have made all the difference. From what paleontologists have been able to tell from bone microstructure, the ancestors and cousins of both crocodiles and dinosaurs grew faster and began reproducing earlier in their lives than proto-mammals. This shorter generation time allowed them to proliferate and be molded by natural selection faster, and they simply outgrew their proto-mammal competition. Mammals didn't have a chance to get big because the faster-growing archosaurs had already taken up that ecological space.

As I learned about the rise of the archosaurs, though, I still had trouble understanding why dinosaurs prospered when similar animals—like *Effigia* and *Postosuchus*—perished at the end of the Triassic. Why didn't the other kinds of archosaurs become the basis for a different, sadly unseen proliferation of extraordinary forms? They were already the dominant form of archosaur on the planet during the Triassic. Dinosaurs were the underdogs.

As the Triassic expert Randall Irmis once explained to me, dinosaurs were few and far between in North America, and no Triassic ornithischians have yet been found on the continent. Even in places they have been found, such as South Africa, ornithischians were small, rare creatures. The only dinosaurs that got big in the Triassic were the primordial precursors of *Apatosaurus*, which are called sauropodomorphs. These dinosaurs were a prominent presence in parts of Late Triassic Europe, but apparently nowhere else. The big picture is that dinosaurs diversified

into a variety of forms soon after their origin but didn't really start
to rule on a global scale until the Jurassic.

Another extinction was the essential event. It's not as infamous as the Permian extinction, and certainly less so than the
mass die-off at the end of the Cretaceous around 66 million years
ago, but there was a distinct drop in the world's faunas at the end
of the Triassic. The usual suspects of climate change, asteroid impact, and intense volcanic activity have all been implicated, without much resolution, but, whatever happened, about half of
the world's biodiversity abruptly disappeared. Many croc-line
archosaurs—such as the phytosaurs, aetosaurs, and rauisuchians—
were among the losses on land, but the disaster doesn't appear to
have affected dinosaurs in the least.

Exactly why dinosaurs outlasted many other archosaur lineages is a question that still confounds paleontologists. What was
it about dinosaurs that let them survive and then thrive while
their cousins perished? A one-size-fits-all hypothesis for the demise
of other archosaurs and the rise of the dinosaurs doesn't exist.

Paleontologists thought they had cracked the puzzle of dinosaur
identity and success. Now we know differently. Serendipity, more
than anything else, gave dinosaurs room to become the creatures we adore. The celebrated story of dinosaurs fighting their
way to the top, crushing all competition underfoot, doesn't work
anymore. "The more new archosaurs we find in the Triassic," according to Sterling, "the less special dinosaurs seem to be."

The end-Permian extinction cleared the way for archosaurs to
diversify thanks to biological quirks such as early reproduction
and fast growth, and the end-Triassic extinction sank other archosaurs that dominated local habitats. As Irmis, Nesbitt, and
other Triassic experts have explained, contingency and oppor-
tunism may have allowed dinosaurs to dominate. Dinosaurs were
awfully lucky. "[T]here was nothing predestined or superior

about dinosaurs when they first arose," Irmis and his colleagues concluded, "and without the contingency of various Earth-history events during the early Mesozoic, the Age of Dinosaurs might have never happened."

What about the evolutionary routes left untraveled? What would the world be like if the end-Permian extinction or the end-Triassic catastrophe had been canceled? Dinosaurs wouldn't have evolved. We probably wouldn't have either, our mammalian forebears forced into the shadows by a different cast of fearsome archosaurs. I like to think of these major events as what fantasy satirist Terry Pratchett once characterized as bifurcations in the trousers of time. The history *we* know went down one leg, but there was another possible outcome. The constrained limits of our imaginations are the only ways to visit those alternate histories. We'll never know all the unrealized evolutionary possibilities, but the survival of the dinosaurs built the foundation for an ongoing dynasty so beautiful and fantastic that I doubt we could have dreamed them if we didn't already know they existed. Even monsters of fantasy—wyverns, the Cyclops, chimeras, and the whole mythological lot—pale in comparison to the real animals paleontologists are continuing to excavate.

For all those stunning forms to exist, though, dinosaurs had to have sex. For vertebrates, at least, that's the process that provides natural selection with raw material to work with—genetic combinations that produce variations which then die off or proliferate according to the constant scrutiny of natural selection. But how do we even approach the question of how *Apatosaurus* made the earth move for each other? We need to look at the intimate details of dinosaur fossils to find out.

Three

Big Bang Theory

There is a dinosaur in Concourse B of United Airlines Terminal One at Chicago's O'Hare International Airport. The first time I saw the skeleton, I thought it was a mirage created by my flight-addled brain.

I hate air travel. I wish I could just opt for the King Kong package—knock me out with sedatives, put me in a crate, and unload me when I get to my destination. My aversion to flying at 30,000 feet isn't helped by my habits. When I travel alone, I am so focused on getting to Point B that I don't stop to eat. Running on the can of soda and tiny pretzel packet from my first flight from the tiny Billings, Montana airport, the first leg on my trip back to my former home in New Jersey after a fossil-hunting trip in Wyoming, I shuffled through the crowded halls to find a seat in the sweltering waiting area. And that's when I saw the dinosaur. I focused on the mirage for a moment, waiting for it to evaporate, but it stayed there—a magnificent, towering skeleton of *Brachiosaurus*.

The fossil cast in Concourse B used to stand in Stanley Field Hall at Chicago's Field Museum, until Sue, the eight-million-dollar *Tyrannosaurus*, came on the scene and bumped the sauropod from its pedestal. In 2000, the sauropod was reassembled among the airport's kiosks and ads. On the day I dragged myself through the concourse, the dinosaur peered over the top of a banner for the airport's Wi-Fi as if it was looking to the tarmac beyond to check the latest departures and arrivals.

This *Brachiosaurus* definitely wasn't the bored slouch I saw in my elementary school's faded textbooks. The heavy shoulders and columnar forelimbs gave the dinosaur a dignified air, only enhanced by the long swerve of neck bones up to a boxy, round-muzzled skull. That skull—and a few other parts—were borrowed from a different, closely related dinosaur from the Jurassic of Tanzania called *Giraffatitan*, but the amalgamation of facsimile bones still reflected the imposing stature of one of the largest dinosaurs ever found. *Brachiosaurus* may not hold the heavyweight title anymore, but for me, an eighty-five-foot-long sauropod is just as impressive as a hundred-foot one. I can only guess how my tiny mammalian ancestors must have seen these giants.

I kept staring at the dinosaur, following the complicated flanges of bone along each vertebra and the rough patches on the limbs where immense muscles would have attached. I let my mind's eye linger and start to fit internal organs into the framework and pull muscle over the whole thing before wrapping the dinosaur in a dappled pattern of large brown spots and white lines—not unlike a giraffe. Not very original, but a little more colorful than the gray-and-olive-green standard I grew up seeing plastered on the dinosaur's hide. And then a strange thought bubbled up from the back of my mind: how did such a gargantuan animal have sex?

Giddy and tired, I envisioned a pair of amorous *Brachiosaurus* standing in a clearing within a Jurassic conifer forest, each waiting for the other to make the first move. I couldn't quite figure out the mechanics of what should come next. I kept looking at the skeleton, as if the fake bones might offer some clue, but all I could think of was that sex must have been damn difficult with that big tail in the way. Alas, my flight started boarding, so I trudged over to take my place among the other tired and impatient passengers. At least the dinosaur mating mystery gave me something to focus

on during my last flight home, and the question has kept nagging
at me ever since.

As the British paleontologist Derek Ager once wrote, "After eat-
ing, the most widespread habits among modern animals are those
concerned with sex, and there is no reason to suppose that this
did not raise its allegedly ugly head millions of years before
Freud." Indeed, sex was the inheritance of dinosaurs and every
other amniote descended from lizard-like creatures that left the
life aquatic behind. As vertebrates took up an entirely terrestrial
lifestyle more than 312 million years ago, sperm could no longer
be squirted onto soft eggs suspended in safe little pools. Tougher-
shelled eggs, harboring an internal pond to nourish developing
embryos, required internal fertilization away from the water's edge,
and so the invention of sex was passed down the generations.
From the tiny, feathery *Anchiornis* all the way up to *Brachiosaurus*
and other gigantic sauropods, dinosaurs were part of this reproductive legacy. "Clearly if we are to regard our fossils as once-
living creatures," Ager wrote, "considerations of sex must arise."

One of the earliest considerations of passionate dinosaur en-
counters was put forward a century ago. In 1906, the American
Museum of Natural History paleontologist Henry Fairfield Osborn pointed to affectionate occasions between fearsome *Tyrannosaurus rex* to explain the dinosaur's oft-ridiculed arms. A pair of
Tyrannosaurus specimens collected by fossil hunter Barnum Brown
unmistakably showed that this dinosaur had short but heavily
muscled forelimbs. Osborn couldn't believe that such small arms
played any role in grappling with big game like *Edmontosaurus* or
Triceratops, but perhaps the arm of *Tyrannosaurus* was "a grasping
organ in copulation." Just imagine two immense predators, one
atop the other and holding on to his mate with those beefy minia-
turized appendages. Sadly, Osborn didn't commission a drawing
of the behavior from the skilled illustrators he often tapped to restore prehistoric creatures.

Osborn didn't give any serious consideration to dinosaur sex, though. Nor did many other paleontologists of his generation. Dinosaur copulation was seen as a silly subject and beyond the reach of investigation. Plus, it seemed to make dignified researchers feel rather squicky. Sex, in natural history, is a perfectly acceptable subject when considering flashy courting behavior or when boiled down into quantitative surveys of gene pools, but the sordid details of sex itself have often made researchers feel awkward. Not long after Osborn briefly mused about *Tyrannosaurus* sex, George Murray Levick—a naturalist with the 1910–13 Scott Antarctic Expedition—was shocked and disgusted by the "sexual depravity" of Adélie penguins (which, you'll recall, we now know are living dinosaurs). He was especially horrified by a young male penguin that tried to mate with a dead female. Levick wrote notes in Greek so that only classically educated scientists like himself would be able to read what he observed, and when he prepared a monograph on the penguins, the passages on sexual behavior were considered so sensational and disgusting that the section was cut and only circulated among a small cadre of scientists. (It wasn't until 2012 that Levick's observations—which were unique for their time—were rediscovered and made publicly available.) Sexual behavior, even among living species, was a taboo subject, and speculating in unseemly detail about the mating habits of dinosaurs would surely highlight a scientist as a pervert. Whatever dinosaurs did on hot Jurassic nights was kept behind the shroud of prehistory.

Not that paleontologists had much to go on. Signs of fossil sex are hard to find. Among the rare examples are 47-million-year-old turtles that died while copulating, and possibly a pair of much older, 320-million-year-old sharks that might have been courting when they were rapidly buried. Sadly, no dinosaur skeletons have ever been found articulated in the act, and not even the most beautifully preserved dinosaurs retain any sign of their reproductive organs. Even evidence of nonsexual organs is exceptionally rare, and sometimes fossils thought to represent internal dinosaur tissue have turned out to be something else.

This was the case in 2000 when the veterinarian Paul Fisher, the paleontologist Dale Russell, and collaborators announced that they had found evidence of a dinosaur heart. The fossil was embedded in the chest of "Willo"—a small herbivorous dinosaur known as *Thescelosaurus*—and appeared to be a four-chambered heart with a single aorta. But in 2011 another team of paleontologists determined that the "heart" was really a blob of iron-rich rock that had cemented together in the position of the dinosaur's chest. They found tidbits of what might have been fossilized dinosaur cells, which hinted that the concretion was created when the contents of the dinosaur's heart were exposed and attracted iron-rich sediment, but the structure itself was not a heart of stone.

Remnants of external dinosaur anatomy might not be of very much help in exploring dinosaur sex, either. Skin impressions are often found in patches, rather than wide swaths covering the whole animal, and even these would not preserve the internal plumbing of dinosaurs. Nevertheless, after reading a 2012 paper on soft-tissue traces preserved with the crested hadrosaur *Saurolophus*, I asked its author, the Philip J. Currie Dinosaur Museum paleontologist Phil Bell, if there might be any possibility of finding soft-tissue traces of the delicate dinosaur bits. After all, I knew that in 2009 paleontologists working on a much, much earlier organism—a 380-million-year-old armored fish called *Incisoscutum ritchiei*—had identified the oldest known fossil evidence of a penis. Maybe there is hope that some lucky paleontologist will have the amazing shock of finding preserved dinosaur copulatory equipment. (I would love to see such a discovery grace the cover of *Science* or *Nature*.)

Bell, however, wasn't optimistic. The trouble with soft tissues, he explained, is that there has to be some way for their presence to be recorded as a fossil. Bell cited cololites as an example. These are different from the infamous coprolites that paleontology professors seem to enjoy handing new students, asking "What do you think this is?" before revealing that the hard lumps are really fossilized dinosaur feces. A *colo*lite is a *premature* coprolite—what Bell described as "fossil dung that hasn't yet been excreted." Gross, but highly informative. Cololites can preserve the form and extent of a dinosaur's intestine, as in a little predatory dinosaur named *Scipionyx* that was found in Italy. In other cases, such as poor Willo, certain organs might be preserved as stains or concretions that formed when the dinosaur's guts began to decay. "What we know of the internal organs," Bell wrote to me, "comes from mineral-rich areas (like the liver) that somehow react favorably in certain conditions for them to be fossilized, or from secondary information (such as cololites) that is more prone to fossilization."

But Bell mentioned one extraordinary fossil that reveals something important about dinosaur reproductive organs. While it's anyone's guess what dinosaurian penises looked like—dinosaur dicks must have been as grotesquely fantastic and terrifying as the rest of their anatomy, right?—paleontologists actually know more about the reproductive organs of female dinosaurs, thanks to a very special pair of hips recovered from Cretaceous China.

Inside the hips of an oviraptorosaur—a feathered, crested, beaked dinosaur that looked something like a terrestrial parrot— the Canadian Museum of Nature paleontologist Tamaki Sato and her colleagues found a pair of preserved eggs. This pregnant dinosaur died just before adding those eggs to her nest. Even better, the pair of eggs showed that this female dinosaur had a combination of traits seen in birds and crocodiles. The presence of two eggs at the same stage of development showed that dinosaurs had paired oviducts, which is a trait seen in crocodylians. But the fact that there were only two eggs indicated that, like birds, the dinosaur probably developed only one egg in each oviduct. This pattern matched what paleontologists saw in the preserved nests of these dinosaurs—the eggs were arranged in pairs, and the gravid dinosaur finally explained why this should be so. The eggs were paired because that's how they were laid.

The unique specimen provided a glimpse of the internal reproductive workings of female dinosaurs. Unfortunately, though, she

did not illustrate the Tab A, Slot B anatomy actually used during
sex. To figure that out, we must turn to their closest relatives, birds.

Even though we usually think of birds as an entirely different
category of creatures, they are truly a specialized lineage of dinosaur that evolved around 150 million years ago and continues to
thrive today. Together with crocodylians—the alligators, gharials,
and crocodiles that are the closest living relatives to the Dinosauria
as a *whole*—birds create a set of evolutionary guideposts called an
extant phylogenetic bracket. The logic is very simple. It basically
means that a trait present in both birds and crocodylians is likely to
have been present in non-avian dinosaurs as well. We can use what
we know about living species to investigate prehistoric creatures.

Take the cloaca, for example. This charming-sounding ori-
fice (from the Latin for "sewer") is the single endpoint for the reproductive, urinary, and intestinal tracts in both sexes of birds
and crocodylians. Following the logic of the extant phylogenetic
bracket, we can deduce that dinosaurs probably had these organs,
too. You wouldn't be able to see anything hanging low, or wob-
bling to and fro, as a male *Apatosaurus* plodded by. The dinosaur's
genitals would be kept inside the cloaca, which would have only
looked like a slit beneath the dinosaur's tail. And since scientists
have been unable to find any definitive skeletal differences between male and female dinosaurs, this means that you wouldn't be
able to tell a female *Allosaurus* from a male one unless you looked
inside the dinosaur's genital opening.

So what did male dinosaurs hide inside their cloaca? Whenever
I bring up the subject of dinosaur sex—which may be too often—
nervous laughter and horrifying speculation about the members
of *Spinosaurus, Brachiosaurus*, and kin soon follow. *Stegosaurus* and
other well-armored dinosaurs always generate the most startling
ideas. Covered in immobile plates and spikes, these dinosaurs
had to avoid being impaled during their dangerous liaisons and
certainly could have used a reproductive work-around. More than
once, my friends have indulged my curiosity, jokingly suggesting

that male dinosaurs had a prehensile phallus that could safely inseminate females from a distance—an idea that shocks my mind into waking anime nightmares. Unfortunately, there's not a scrap of fossil evidence to say whether such terrifying organs actually existed.

We have to go back to birds and crocodiles for clues. Birds are a highly varied group of animals with different reproductive arrangements and strategies. While most male birds don't have a penis at all—they pass semen to females by way of a technique given the cringe-inducing name "cloacal kiss"—some birds are impressively endowed. The Argentine lake duck *Oxyura vittata* has the longest penis relative to its body length of any known vertebrate. According to the ornithologist Kevin McCracken, who actually went to the trouble of investigating the bird's infamous appendage, the length of the duck's spine-covered penis can rival that of the bird's body. Other ducks, male and female, are similarly famous for having prominent, complex sexual organs that fit together like a key in a lock or a limber corkscrew in a labyrinth—flashy outcomes of a sexual evolutionary arms race between males and females. And while not all male birds have penises, the lineages with well-endowed males are all near the base of the bird family tree. As determined by the University of Massachusetts (Amherst) ornithologist Patricia Brennan and her colleagues, this means that a penis is the ancestral state for birds.

Like waterfowl, male alligators, crocodiles, and gharials have penises, too. In their survey of crocodylian genitals, zoologists Thomas Ziegler and Sven Olbort wrote: "For correct sex identification, the male external genital organ of an immobilized crocodile must be felt out and protruded from the cloaca, and subsequently compared to the similar female clitoris." Not an enviable task. But together the "intromittent organs" of crocs and birds indicate that male dinosaurs almost certainly had penises. And even though we don't know exactly what these organs looked like, both crocodylians and endowed birds have a single, unpaired organ with at least one long runnel down which semen flows during sex.

How dinosaurs varied from this basic structure is unknown. I doubt there was a one-size-fits-all appendage. In 2006, Steve Wang and Peter Dodson estimated that more than 1,850 genera of non-avian dinosaurs lived during a span of over 150 million years. With so much diversity, there must have been startling variations of penis structure and female variations to match. (Sadly, finding a dinosaur clitoris seems as unlikely as finding a fossil penis. Some problems are eternal.)

Reconstructing the mating habits of dinosaurs requires a bit more than a basic understanding of the anatomical tackle they used, however. We need to know which dinosaurs were males and which were females, and, more important, if there were any broader physical differences between the sexes. Among living birds and crocs, the cloaca is the key to sex identification. Paleontologists obviously don't have that option. So, in lieu of soft tissues, and given the rarity of obviously female skeletons with eggs preserved inside, paleontologists searched for skeletal characteristics that might distinguish one sex from another. Until recently, the key was thought to be sexual dimorphism—differences between the sexes other than the anatomy of the genitals—including size, ornamentation, and other features that might distinguish males from females.

There's only one problem: identifying sexual dimorphism requires a large sample size from a population of dinosaurs that lived at a particular place and time. Paleontologists rarely get such detailed samples. Even our knowledge of *Tyrannosaurus*, arguably the best-studied dinosaur, comes from about fifty specimens spanning 2 million years. Unless a large number of animals died together in the same place at the same time, we're out of luck.

These frustrating circumstances haven't stopped paleontologists from trying. A study done back in 1975 by the paleontologist Peter Dodson highlighted the trouble with distinguishing male and female dinosaurs. During the previous hundred years, paleontologists had named three genera and twelve different species of crested hadrosaur from the roughly 74- to 76-million-year-old strata of Alberta, Canada's Oldman Formation. When Dodson compared measurements of the dinosaurs' skulls, however, he found only three species of hadrosaur present in the same place at roughly the same time—*Corythosaurus casuarius, Lambeosaurus lambei,* and *Lambeosaurus magnicristatus.* Each of these dinosaurs had a similar body shape, but they differed in their crests—*Corythosaurus* had a low, rounded crest, while the *Lambeosaurus* species had hatchet-shaped and large, curved crests, respectively. What paleontologists had believed to be species differences were just variations from one individual to the next, or represented age stages of the same species. But Dodson also believed that he had uncovered sexual dimorphism in each of the three valid species he proposed. In *Lambeosaurus magnicristatus,* for example, one sex had a bulging pompadour of a crest while the other seemed to have a more abbreviated version of the same structure. Dodson suspected that the more prominently crested form was a male and the poorly endowed form was a female.

Other paleontologists even went so far as to suggest that the two species of *Lambeosaurus* should really be one, with the larger-crested forms being males and the smaller-crested types, females. After all, crests are very prominent visual signals, and male *Lambeosaurus* might have flashed their decorations as signs of dominance, vigor, and power. The larger the crest, the more desirable the male.

But these paleontologists had misread the evidence. In their analysis of some of the fossils Dodson studied, the Canadian paleontologists David Evans and Robert Reisz found that *Lambeosaurus magnicristatus* truly was a separate species, not just a male of another. Not only did this species live at a different time than the other *Lambeosaurus* species, but the skull Dodson designated "female" had been broken so that it only appeared smaller. There was no evidence for sexual dimorphism in these dinosaurs.

Even *Tyrannosaurus rex* suffered from some sex-identity confusion. During the 1990s, some paleontologists suggested that female

T. rex had a wider space between the base of their tail and their hips to allow eggs to pass through. If true, this would be a way to identify female dinosaurs, but the argument crumbled. As paleontologists later found by studying fossils and the anatomical differences between male and female alligators, the proposed gap didn't actually exist.

Other supposed indicators of dinosaur sex have not fared well, either. Sexual skeletal differences have been proposed for a variety of dinosaurs, from the sleek little early theropod *Megapnosaurus* to Mongolia's famous ceratopsian, *Protoceratops*, and the super-spiky stegosaur *Kentrosaurus*, but none of the traits in these dinosaurs has turned out to be a reliable sex indicator. While dinosaur sample sizes are small, and cases of sexual dimorphism may be discovered one day, no one has yet found an unequivocal case.

Since skeletal differences between male and female dinosaurs are so elusive, if there are any at all, the only way that we can identify dinosaur sexes is through more direct evidence. Finding developing eggs inside a dinosaur's body cavity—as with the rare oviraptorosaur specimen—is one way to pinpoint a female dinosaur. But there is another option.

In 2000, a special specimen of *T. rex* finally yielded a way to unmask female dinosaurs. The Museum of the Rockies field crew, directed by the museum's curator of paleontology, Jack Horner of Montana State University, hit the fossil jackpot: in the course of a summer excavating Montana's Hell Creek Formation, they found five *Tyrannosaurus* specimens, including one nicknamed "B. rex" after its discoverer, the crew chief Bob Harmon. The position of this dinosaur twenty feet up a steep cliff made the excavation process especially difficult, and when the skeleton was finally dug out, the dinosaur's jacketed femur was too heavy for a hired helicopter to fly out of the remote field site. The paleontologists reluctantly broke the thigh bone in two for transport. This decision ended up making all the difference.

The frustrating field circumstance turned into a lucky break for one of Horner's former students, Mary Schweitzer. Schweitzer is an expert in the microscropic study of dinosaur tissues, and at the time of the B. rex discovery, she was searching for fossils that hadn't been doused with Vinac, glue, and the various other adhesives paleontologists apply to keep fossils together in the field. Fresh, untampered-with fossil surfaces had the potential to reveal some aspects of dinosaur biochemistry. Fortuitously, Horner sent Schweitzer, who had just started teaching at North Carolina State, some fragments that had fallen from the freshly broken femur. The gift offered her an opportunity to carefully analyze the unadulterated fossil bone of *T. rex*.

Schweitzer was shocked by what she found. The *Tyrannosaurus* had been pregnant when it died. A peculiar type of bone tissue inside the femur gave the girl away. When some species of female birds are growing eggs, they develop a thin layer of tissue called medullary bone inside the shafts of the long bones in their hind limbs. This tissue is calcium-rich and acts as a store of raw material for creating eggshells, and it was what Schweitzer observed in B. rex. Not only did the discovery mean that this physiological response to pregnancy evolved in non-avian dinosaurs first— *Tyrannosaurus* was a distant cousin of early birds, and so the characteristic must have evolved deeper within the dinosaur family tree—but Schweitzer and her collaborator Jennifer Wittmeyer discovered a subtle way to identify some female dinosaurs. The technique wouldn't work for all females—the dinosaurian hens would have to be laying eggs for the tissue to form—but medullary bone was still a way to pick at least a few females out of the vast collection of dinosaurs that paleontologists have accumulated.

Graduate students at the University of California, Berkeley, Andrew Lee and Sarah Werning ran with this discovery to investigate the tempo of dinosaur sex lives. While the discovery of medullary bone was new, paleontologists had been tracking dinosaur growth through bone microstructure for years. Previous research had established that dinosaurs have rings in their bones which can be used to estimate the age at which they died. These bands, called lines of arrested growth (or simply referred to as LAGs), most likely represent a regular slowdown in growth during especially tough times, such as a dry season when water and food are scarce. This works to the benefit of paleontologists. These rings, combined with studies of bone tissue types and reconstructions of dinosaur growth curves, indicated that many dinosaurs grew very rapidly during their early lives, and eventually grew more slowly as they approached skeletal maturity.

In addition to investigating B. rex, Lee and Werning found traces of medullary bone in two other kinds of dinosaurs—a beaked Early Cretaceous herbivore called *Tenontosaurus* and the Jurassic carnivore *Allosaurus*. All the dinosaurs were young moms. Lee and Werning estimated that the *Tenontosaurus* died at eight years of age, the *Allosaurus* at ten, and the *Tyrannosaurus* at eighteen. All of these dinosaurs were still growing when they died—their skeletons had not yet developed to full maturity. Based on these findings, it seemed that dinosaurs in general might have started reproducing even earlier. Regarding each of the dinosaurs in the study, the cases of medullary bone in each indicated only the latest date at which each kind of dinosaur started having sex.

Dinosaurs lived fast and died young. Rapid growth and early reproduction, Lee and Werning suggested, might be a sign of difficult, dangerous lives in which mating early was required for a dinosaur to pass along its genes to the next generation. This would have been especially important for the biggest dinosaurs. If an eighty-foot dinosaur such as *Apatosaurus* took decades to grow to sexual maturity, there would be very few of them around to breed. Instead, Lee and Werning estimated, these dinosaurs probably started copulating long before they reached maximum size, probably by nineteen years of age. Teenagers will be teenagers, after all.

Paleontologists are starting to outline the biology of dinosaur sex. But the mating behavior of dinosaurs remains mysterious.

How did dinosaurs pick their mates and show off? Did brightly colored *Ceratosaurus* strut through the mating grounds like John Travolta? Did *Apatosaurus* nuzzle each other's neck to give the proceedings a sense of romance? Paleontologists have had their fun with this topic when they have waded into paleo-fiction. In his children's book *The Year of the Dinosaur*, the late paleontologist Edwin Colbert avoided detail to present only a perfunctory view of dinosaur courtship among a group of *"Brontosaurus."* "[A]s their desires reached a peak, copulation took place," Colbert blandly described. Not exactly the most tender affair. Although, in artist William Stout's lavishly illustrated *The Dinosaurs*, William Service added a little more color to the proceedings and imagined a *Parasaurolophus* breeding ground where males uttered sonorous notes "[r]icher, deeper, thicker than any other animal tone heard in the air before or since" in their attempts to attract females to their little patches of wetland.

Whether dinosaurs were fans of love bites or preferred to cuddle is forever lost to time. Nevertheless, paleontologists have wondered whether many of the bizarre features of dinosaurs might be related to their sex lives. The connection between dinosaur ornamentation and mating is an application of one of Charles Darwin's most important theoretical frameworks—sexual selection. Creatures not only struggle to survive, Darwin recognized, but also to reproduce. In some mating systems, males compete with other males of their species for access to fertile females. Females also have different interests than the males, especially since they often take on the brunt of reproductive responsibility: a male has to invest only a little sperm, while a female contributes far more resources to a developing embryo. And this is to say nothing of conflicts among females themselves. The competition within and between the sexes can produce complicated signals and bluffs, from the artistic architecture employed by male bowerbirds to the deep-throated roars and powerful displays of American alligators during the mating season. It all comes down to sex. The best package of survival traits is meaningless if the individual animal in possession of those characteristics never breeds.

Since we can't observe the behavior of non-avian dinosaurs directly, their skeletons are all we have to gauge their sex lives. Crests, spikes, plates, horns, and feathers can all act as prominent display structures, and maybe such ornaments acted as indicators of a dinosaur's health or dominance. As the paleontologist and popular host of the kids' show *Dinosaur Train* Scott Sampson has suggested, the strange adornments were so dazzling that even we cherish dinosaurs for them—sexy structures enthrall us just as they may have attracted amorous dinosaurs.

But picking out the influence of sexual selection in the fossil record is extremely difficult, especially when its most classic sign— sexual dimorphism—is absent among dinosaurs. The long, elegant necks of *Apatosaurus*, *Brachiosaurus*, and their sauropod kin were recently at the center of a debate over whether their strange shapes betrayed the influence of sexual selection. Giraffes inspired the hypothesis.

The classic story of how the giraffe got its neck is that the gradual extension of the mammal's cervical vertebrae allowed the herbivore to reach food above the heads of competitors. But in 1996 the zoologists Robert Simmons and Lue Scheepers proposed something different. Male giraffes frequently fight each other in competitions called "necking." It's not as nice as it sounds. The fighting males swing their long necks to batter each other with the stout ossicones on the tops of their heads. These contests establish dominance; presumably, more dominant males that control territories end up with more mating opportunities than the losers (provided the losers survive the sometimes lethal ordeal). Therefore, Simmons and Scheepers proposed, such competitions drove the evolution of giraffe necks. Starting with a relatively short-necked ancestor, males who had necks just a little bit longer—and had a more powerful swing—would win more often and therefore mate more frequently. Over time, this selection for longer and longer necks would result in the giraffe as we know it today. Their idea became known as the "necks for sex" hypothesis.

A decade later, the Fayetteville State University paleontologist Phil Senter applied the same idea to sauropods. Senter argued that long sauropod necks did not give dinosaurs any appreciable vertical reach, but instead only helped the dinosaurs browse on low-lying plants. Elongated necks, in Senter's view, also might have made dinosaurs conspicuously vulnerable to attack from Jurassic predators. All an *Allosaurus* or *Torvosaurus* would have to do, he suggested, was chomp into an exposed *Apatosaurus* neck to gain many tons of succulent sauropod meat. Since sauropod necks seemed to have little utility in helping their owners survive and may have even come at a high cost, Senter concluded, the long necks were probably visual signals that had somehow become established through generations of mating.

But the sauropod experts Mike Taylor, Dave Hone, Matt Wedel, and Darren Naish rebutted Senter's sexual selection argument. Sauropods probably kept their necks elevated much of the time, the paleontologists pointed out, and this gave the dinosaurs a distinct feeding advantage—an *Apatosaurus* could stand in one place and browse up, down, and from side to side without having to move. (Giraffes, too, have been shown to gain a feeding advantage from their neck, overturning the "necks for sex" hypothesis.) And the graceful necks of sauropods—supported by ligaments, tendons, and bone—were actually quite sturdy parts of the body that an attacking *Allosaurus* would have a hard time biting into. The four argued that, contrary to Senter's assessment, sauropod necks were sturdy organs essential to the success of the giants.

That's not to say that the necks of *Apatosaurus* and kin had nothing to do with mating. Sauropod necks likely evolved to allow the dinosaurs to reach a wide range of food, after all, but their necks could have been co-opted for sexier purposes. During the mating season, sauropod necks might have acted as huge fleshy billboards advertising individuals' fitness. Since many of these giants were too big to worry about hiding—predatory dinosaurs probably targeted juvenile and small individuals over the biggest behemoths—immense sauropods could dispense with camouflage. Might the necks of dinosaurs like *Brachiosaurus* have been especially colorful during the mating season, ornamented with striking color patterns to gain the attention of potential mates and show off that they were the healthiest, most desirable dinosaurs around? These are the kinds of questions that can keep a paleontologist up at night.

Other dinosaurs may have shown off a bit, too. Maybe the spiky dinosaur *Kentrosaurus* found the plates and spikes of the opposite sex arousing, and perhaps females of the sauropod *Amargasaurus* looked for males with the longest neck spines. We don't know for sure, but such prominent visual signals probably had some utility in the mating game.

Ornamentation can be a way for potential breeding partners to detect if there's truth in advertising. Magnificence costs energy, which the individual flaunting it must have to spare. A good mate will look healthy and especially flashy. And there was probably a lot of stomping, hooting, and showing off during the Mesozoic when dinosaurs tried to impress each other. But this only leads us right back to that most persistent of dinosaur sex mysteries: after all the posturing and showing off, *how did they do it?*

Let's assume that male dinosaurs didn't have monstrous penises. (My apologies to Mr. Rex.) Mating dinosaurs would have had to bring their cloacae into close contact for sex to work. How this actually happened depends on what dinosaurs were physically capable of, and on that score we can only speculate. In 1993, the American Museum of Natural History unveiled a stunning Jurassic vignette: a mother *Barosaurus* rearing up to defend her offspring from an *Allosaurus* about to pounce on the youngster. The museum made it clear that the scene was speculative—a vision of what dinosaurs might be able to do, rather than what we know they did—but the reconstruction caused a stir among paleontolo-

The reconstructed skeleton of the Jurassic stegosaur *Kentrosaurus.* Did these dinosaurs use their spikes as sexy signals? Even if not, how did they go about the delicate
business of mating? (Photograph by H. Zell, from Wikipedia: en.wikipedia.org/wiki/File:Kentrosaurus_aethiopicus
_01.jpg)

gists. How could an animal so immense rear back without snap-
ping its hind limbs like toothpicks? And how did the dinosaur's
heart manage to pump enough blood up to the sauropod's head?
Traditionalists saw the mount as unbridled, sensationalistic speculation, while advocates of dinosaurs as agile, highly active animals viewed the display as a perfect example of this new view.

When the famous exhibit came up in my Paleontology 101
course at Rutgers University, my professor pointed out that some
sauropods must have reared up in just such a position every now
and then. How else would males be able to mate with females?
The biomechanics expert R. McNeill Alexander made a similar
point. Alexander imagined that dinosaurs mated just like today's
elephants and rhinos: females had to bear the extra weight of the
mounting male. The main difference would be that dinosaurs had
those big, relatively stiff tails. Working from the idea that a male
dinosaur threw one of his forelegs over the back of the female,
Alexander pointed out that the weight of the male would have
rested on the female's hindquarters. This would have been a massive load, but, Alexander noted, the stresses involved wouldn't
have been any worse than walking, since, during the step cycle, the
dinosaur's own weight would have been supported by just one hind

leg as the other swung through the air. "If dinosaurs were strong enough to walk they were also strong enough to copulate," Alexander wrote. "They were presumably strong enough to do both."

All the same, what positions were in the dinosaur playbook has remained in the realm of speculation. When I described the problem to my wife, who thankfully doesn't think it at all strange to wonder about dinosaur sex lives, Tracey replied, "You'd think the genitals of the females would be in a more convenient place. Maybe on the side, like a gas tank." Clearly, dinosaurs are yet another example of unintelligent design.

However they did it, though, it's clear that male dinosaurs had to somehow mount female dinosaurs. That's the general idea the paleontologist Beverly Halstead ran with when he pontificated on dinosaur sex for live audiences. (He is a legend among paleontologists for regularly bringing his wife on stage during lectures to demonstrate positions from the Mesozoic Kama Sutra.) He believed that dinosaurs did as lizards and alligators do today. Male dinosaurs grasped or leaned on the female with their forelimbs and threw one hind leg over her back, Halstead surmised, and this would push the male's hips beneath the tail of the female to bring their cloacae into contact. Longer-tailed species may have even intertwined their tails, just as some snakes twist their bodies around one another. Halstead's version of dinosaur style, with a few variations, became the favored hypothesis.

I've never been satisfied by this standard explanation of dinosaur sex. It's easy to draw a two-dimensional image of dinosaurs mating, but we don't really know if their legs and tails could have bent and flexed enough to achieve the traditional position. And the mating habits of stegosaurs defy the imagination. Just thinking of how *Kentrosaurus* managed gives me a headache. This relative of the more famous *Stegosaurus* had small plates armoring its neck and upper back that transitioned rearward into paired sets of huge spikes, including large weapons over the hips. The leg-over-back position just wouldn't work—I know because I asked my paleontologist friend Heinrich Mallison to check.

Using the *Kentrosaurus* mount at the Museum für Naturkunde in Berlin, Mallison had previously created a digital scan of the dinosaur's skeleton to investigate just how flexible this spiky herbivore was. Among other things, Mallison discovered that *Kentrosaurus* could swing its tail in a seventy-five-degree arc at an estimated ten meters per second. "At this speed," he wrote, "the spikes could penetrate deeply into soft tissues or between ribs and were able to shatter bones." Not a dinosaur you'd want to piss off. But reading his papers made me think about the more tender moments in the life of *Kentrosaurus*. If Mallison had made flexible virtual models of the dinosaurs, then those models could be used to test dinosaur sex positions in three dimensions.

Suggesting that a friend and researcher make his carefully assembled virtual dinosaur models have pixelated sex is a delicate proposition, but thankfully Mallison was enthusiastic about the idea. And the traditional dinosaur sex position didn't work. If a male *Kentrosaurus* tried to throw his leg over the back of a crouching female, he'd castrate himself on her sharp spikes. One hip spike, in particular, seemed to be positioned just right to strike fear into the hearts of stegosaur suitors. More than that, the tails and hips of these dinosaurs so strictly limited their mobility that the classic dinosaur-style position was impossible. These prickly dinosaurs must have done something else, and, with luck, making virtual dinosaurs fool around might help solve the problem.

One of the most essential aspects of dinosaur lives remains modestly cloaked in mystery. Still, *Kentrosaurus* and kin obviously figured it out well enough to continue their kind. In fact, the products of their prehistoric unions—eggs and baby dinosaurs—are beginning to show paleontologists just how dramatically dinosaurs changed as they grew up.

Four

The Dinosaurs,
They Are a-Changin'

A long time ago, in elementary school science class, my peers and I watched a baby chick peck its way out of an egg. At the time, I had no idea that I was seeing a dinosaur being born. To my younger self, birds and dinosaurs were entirely separate branches on the tree of life. Now I know better. Looking back on that little theropod, it's clear that the chick was carrying on the dinosaur legacy. Pick any dinosaur you like—all the cute infants greeted the world by breaking their way out of an egg.

The simple fact that dinosaurs laid eggs took decades to confirm. For a full century after the first dinosaurs were discovered, paleontologists didn't really know how the creatures gave birth to the next generation. Some, such as William Diller Matthew, entertained the idea that dinosaurs were capable of live birth; perhaps pregnant females carried one large offspring at a time, like elephants. Others thought dinosaurs were more like modern reptiles, packing nests with eggs. It wasn't until the 1920s, when an American Museum of Natural History expedition to Mongolia returned to the United States with undeniable dinosaur eggs, that paleontologists confirmed that baby dinosaurs hatched in carefully constructed nests. This raised additional questions. Did little dinosaurs start off their lives as miniature copies of adults, and did their parents care for them at all? These questions have surrounded subsequent discoveries of dinosaur eggs, nests, babies, and adults fossilized in brooding positions. By probing these intricate specimens, paleontologists found something unexpected. Dinosaurs didn't just look weird—they grew up in a unique way.

In fact, the eggs the AMNH team discovered in Mongolia were even more important than paleontologists understood at the time. The oval fossils were not only evidence that dinosaurs constructed nests. Paired with a later, fortuitous find, those eggs have revealed a dramatic dinosaurian irony that further sets the scene for how baby dinosaurs started life.

When the AMNH expedition collected the eggs from the Gobi Desert, paleontologists thought they belonged to a small ceratopsian dinosaur named *Protoceratops* that they found nearby. But the same crew also found another dinosaur associated with a Cretaceous nest—a graceful, birdlike theropod dinosaur with a toothless beak. When he described the animal in 1924, Henry Fairfield Osborn called it *Oviraptor philoceratops*, an "egg seizer" with a "fondness for ceratopsian eggs," because the dinosaur's skull was found right on top of a clutch of them. The close juxtaposition of *Oviraptor* and the nest, Osborn wrote, "put the animal under suspicion of having been overtaken by a sandstorm in the very act of robbing the dinosaur egg nest." The supposed predator's red-handed shame was immortalized in its moniker.

Only, the *Oviraptor* wasn't stealing the eggs. We know thanks to a beautiful baby dinosaur that was uncovered much later, in 1994. The AMNH paleontologist Mark Norell and his colleagues described an astounding, delicate fossil they discovered the year before at another fossil-rich site in the Gobi Desert. Curled up inside an egg was the embryo of an oviraptorid dinosaur—the poor baby was so well developed, in fact, that the paleontologists expected that it was close to hatching when it was entombed in sediment. And the anatomy of the egg that encompassed the little guy was the same as all the so-called *Protoceratops* eggs the original Mongolian expedition had collected so many years before. Astonishingly, Osborn's *Oviraptor* was really protecting her own nest. Due to the rules of taxonomy, though, her kind will always be saddled with the "egg seizer" name.

At first, the idea of a nurturing *Oviraptor* parent was based on the circumstantial evidence of corresponding egg shapes. But soon after the embryonic dinosaur was found, Norell and other paleontologists documented oviraptorid skeletons preserved brooding on top of their nests. These dinosaurs, which were covered in gorgeous plumage, spread their feathery arms over clutches of eggs.

Oviraptor wasn't the only dinosaur to watch over its shell-bound offspring. In fact, a major discovery that had been made years before and a continent away had revealed that dinosaurs were attentive to their nests, and even to hatchlings. During the 1970s, following a tip from fossil hunter Marion Brandvold, the paleontologists Jack Horner and Bob Makela discovered a vast nesting ground of hadrosaurs they called "Egg Mountain" in Montana. This was a fossil bonanza. Nests, eggs, and dinosaurs from infants to adults were all found in the same place. Even better, the anatomy of the youngest dinosaurs showed that they had hatched only a short while before but were not yet strong enough to leave the nest. The hadrosaur babies must have stayed in the nest for at least a short time, the researchers argued, and would have relied on their parents for food. No wonder Horner and Makela decided to name this hadrosaur *Maiasaura*—the "caring mother lizard."

Around the same time that Makela and Horner found Egg Mountain, yet another discovery of a nesting ground on another continent showed that the *Maiasaura* site wasn't a fluke. Many different kinds of dinosaurs cared for their young. In 1976, the paleontologist James Kitching discovered a cache of dinosaur eggs in roughly 190-million-year-old strata of South Africa's Golden Gate Highlands National Park. Kitching presumed that the eggs belonged to *Massospondylus*—a lanky, bipedal dinosaur with a long neck, small head, and short arms tipped in large claws, which had been found in rock of about the same age. (This was the archetype from which the later sauropods would evolve.) What Kitching didn't know was that some of those eggs contained very rare preserved dinosaur embryos. In 2010, after the eggs had been fully cleaned and prepared, the paleontologist Robert Reisz and colleagues found that two of the eggs contained the bones of tiny sauropodomorphs. The nascent dinosaurs were awkward little things. With their short necks, stubby little legs, and big eyes, they were unabashedly cute, and, while different from that of the adults, their anatomy matched what Kitching had proposed.

That wasn't all. When Reisz and colleagues relocated the fossil site, they not only found eggs, but nests and footprints spread through multiple rock layers. This was a place where mother *Massospondylus* returned year after year to nest, and the tracks of baby dinosaurs indicated that the infants remained at the nesting site after hatching—at least until they doubled in size. Since the babies stayed in the nest, chances are that one or both parents stayed to care for them. Birds protect their nests, and modern crocodylians guard their young for a short time after birth, so it's reasonable to suppose (remember the principle of the extant phylogenetic bracket?) that *Massospondylus* did the same.

Dinosaurs didn't just pick any spot to make their nests. Some species even looked for breeding grounds with special perks. One of the most fantastic dinosaur nesting grounds discovered, in the Early Cretaceous rock of Argentina, was a basin filled with hot springs and other geothermal features where sauropods made nests that were warmed by the natural heat emanating from the earth. Just imagine the long-necked leviathans walking through a prehistoric version of Yellowstone's otherworldly geyser basins.

Whether these dinosaurs cared for their eggs or just left them to incubate in the heated nests is unknown. Some paleontologists think that sauropods may have employed a "lay 'em and leave 'em" strategy—that the huge dinosaurs excavated nests, deposited their offspring, and moved on about their business.

How much parental care dinosaurs provided seems to have differed from species to species. Indeed, for dinosaurs other than sauropods, there are tantalizing clues that adults and their young stayed together for extended periods of time. In southwestern Montana, for example, paleontologists have found burrows made by *Oryctodromeus*—a small, bipedal, beaked ornithischian dinosaur. We know this dinosaur species made the burrows because their skeletons have been found inside the holes they created, including remains of an adult and two juveniles preserved within a large burrow chamber. These unfortunate dinosaurs were buried together at the same time, suggesting that some fast-growing dinosaur kids stayed with their parents for months or even years after hatching.

Sooner or later, just like other species, most dinosaur young parted ways with their parents. According to David Varricchio of Montana State University, one of the researchers who studied the *Oryctodromeus* den, who has attempted to put together a generalized picture of dinosaur life history, multiple discoveries of dinosaur bonebeds—from *Triceratops* to the sauropod *Alamosaurus* and the superficially ostrich-like theropod *Sinornithomimus*—that contain juvenile animals of the same age suggest that young dinosaurs hung out together after leaving the nest. These immature dinosaurs were big enough to fend for themselves, but didn't go off on their own or join more mature individuals of their own kind. Instead, many young dinosaurs led a separate existence from adults of their species, and often looked like the awkward teenagers they were.

Through the ongoing investigations of embryos, infants, and juveniles, paleontologists have discovered a rule that is consistent across the non-avian dinosaur family tree: baby dinosaurs didn't look just like miniature copies of their parents. Infant *Maiasaura* were big-eyed, short-snouted babies that would undoubtedly be a hit on websites like Cute Overload if they were around today, and the six-inch *Massospondylus* babies that toddled around the nesting grounds on all fours hardly looked anything like their twenty-foot-long, bipedal parents. Sweeping anatomical changes transformed gawky infant dinosaurs into impressive adults, and the alterations were so drastic that paleontologists have sometimes mistaken juvenile dinosaurs of an already known species for an entirely new kind of dinosaur. This problem has become a growing point of contention among paleontologists, and no beast represents this debate better than one of the greatest dinosaur ambassadors to the public—*Triceratops*.

In 2010, ill-informed news services wailed that Jack Horner and his PhD student John Scannella had banished *Triceratops* into the dustbin of discarded dinosaurs, where "*Deinodon,*" "*Trachodon,*" and many others had come to rest. Some journalists woefully misconstrued what was actually happening, missing the point that many dinosaurs went through elaborate growing pains.

It all started in July of that year, when Scannella and Horner published a paper on how the appearance of *Triceratops* changed as the dinosaur aged. Previous fossil finds and research conducted by Horner and collaborator Mark Goodwin had documented how the dinosaur's horns and frill transformed from infancy to adulthood, but Scannella and Horner found something else. They realized that the specimens they'd thought were adult *Triceratops*—big-headed herbivores with forward-pointing brow horns and a solid frill—hadn't finished growing. All the large *Triceratops* were really young adults, and the fully mature adults had been misidentified as a different dinosaur genus.

This tension between dinosaur biology and our efforts to name and classify dinosaurs is as old as paleontology itself. *Triceratops* is a classic case—our image of this dinosaur has been shifting ever since naturalists first discovered its remains in 1887, at the height of the Bone Wars between the dueling paleontologists E. D. Cope and O. C. Marsh. It was in that year that George Cannon, a high school teacher and amateur geologist, discovered a pair of large horns and part of a skull roof in an exposure near Denver, Colorado. Cannon sent the horns and a few additional fragments to Marsh in New Haven, and, after examining them, the Yale paleontologist believed that the weapons must have belonged to some

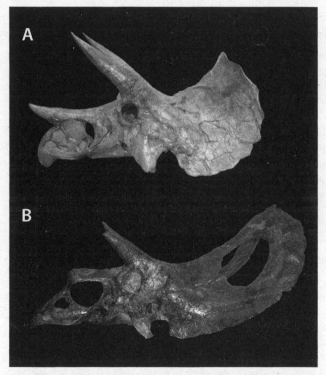

The skulls of *Triceratops* (A) and *Torosaurus* (B). Were these truly two different dinosaurs, or was *Torosaurus* the fully mature form of *Triceratops*? (Illustration from www.plosone.org/article/info%3Adoi%2F10.1371%2Fjournal .pone.0032623)

enormous herbivore. The bones seemed to resemble the horns of America's classic Western icon, the bison, and so Marsh dubbed the mysterious animal *Bison alticornis*.

But there was something amiss about Marsh's bison. Cannon had found definite dinosaur bones in the same layer as the horns. Why would a bison be found in the same stratum as Cretaceous dinosaurs? The puzzle wasn't resolved until Marsh received a partial dinosaur skull from another fossil hunter in 1889. The horns of this new animal were similar to those of the giant "bison," and so Marsh reasoned that Cannon's creature must have also been a dinosaur. With this new data in hand, Marsh recast his "bison" as the dinosaur *Triceratops horridus* the same year, and drew from a different specimen to name a second species in 1890, which he called *Triceratops prorsus*.

Now, as we saw during the "*Brontosaurus*" episode, Marsh had a habit of naming new species from partial remains that differed only slightly from other known fossils. A different curve of horn or an altered angle of armor plate was all that Marsh felt he needed to establish a new species or genus. No surprise, then, that when Marsh's assistant John Bell Hatcher recovered a pair of unusual partial skulls from the *Triceratops*-bearing beds of Wyoming in 1891, Marsh thought the fossils represented a new ceratopsian genus he called *Torosaurus*. *Torosaurus* was another three-horned dinosaur, similar to *Triceratops*, but this big-headed herbivore had an elongated frill with two large circular holes in the parietal bones.

For more than a century, paleontologists followed what Marsh had determined. When Scannella and Horner reexamined the horned dinosaurs, though, they found *Torosaurus* wasn't really a distinct dinosaur. Marsh's *Torosaurus latus* lived in the same place and at the same time as *Triceratops*, and other paleontologists had noted how the two dinosaurs could be told apart only on the basis of a few skull characteristics. These two lines of evidence revealed that *Torosaurus*, rather than being a unique dinosaur, was actually the fully mature form of *Triceratops*. Scannella and Horner argued that late in the dinosaur's life, the frill expanded, large holes opened up in the parietal bones, and triangular ornaments (or epiossifications) skirting the edge of the structure split and flattened.

This dramatic osteological transformation is laid out in full at the Museum of the Rockies in Bozeman, Montana. I had the chance to stop by during a break from fieldwork in the summer of 2011. A prowling *Tyrannosaurus* nicknamed Big Mike greets visitors to the primarily paleontological institution, where authentic fossils, sculptures, and casts all find their place among the dimly lit corridors. As I stroll through, I stop to admire the skull of an *Allosaurus* named "Big Al" and examine an exploded *Tenontosaurus* skull that shows each individual bone sitting in a glass case near a life-size recreation of the herbivorous dinosaur suffering under the switchblade claws of a feathery *Deinonychus*. Both dinosaurs are cleft in two—fully fleshed-out restoration on one side, bare bones on the other.

The huge ceratopsian in the last room is displayed the same way. Head down and horns angled forward, the enormous dinosaur is only half dressed; he has skin on one side and an X-ray view on the other. I want to call the dinosaur *Torosaurus* because of the gaping hole in the frill, but the neighboring display places this aged specimen at the very end of the *Triceratops* growth series—an array of authentic fossils and casts that record the dinosaur's life from infancy to old age. In this case, *Triceratops* skulls, all in a row, display almost the entire life of the dinosaur. A tiny, short-frilled infant with only nubs for horns starts the parade, followed by increasingly larger skulls with long brow horns that curve backward and then come forward again as the dinosaur ages. The biggest skull, a copy of the one on the full sculpture, seems to slot right into the series. With a blunted nose horn, forward-curved brow horns, small epiossifications, and two holes in the frill, this creature would seem to be the elusive final form of *Triceratops*. I consider the displays at length, and they're lovely, but why did *Triceratops* change so much? And where are the specimens that document that last, important change between the solid-frilled type and the one—exemplified by *Torosaurus* specimens—with perforated headgear? I decide to call on Mark Goodwin, the University of California, Berkeley, paleontologist who had worked with Horner to describe how baby *Triceratops* grew into burly adults.

Berkeley's Valley Life Sciences Building, where Goodwin works, houses the university's museum. This is a research institution rather than a gallery of public displays, but there are still a few reconstructions along the building's corridors. A cast of our prehistoric forerunner "Lucy"—the famous *Australopithecus afarensis* skeleton described by Berkeley's anthropologist Tim White—lies prostrate beneath a tombstone reading "R.I.P. Lucy, 3,200,00," and an impressive cast of a *Tyrannosaurus* nicknamed "Wankel rex" snarls at undergrads as they pass him on the staircase to the building's library.

If you visit the library, you'll be greeted by a pair of *Triceratops* skulls. The larger of the two looks like the typical image of the dinosaur—long nose, three horns, solid frill—but the smaller specimen, held up on a little stand, isn't any bigger than my own skull. The cast is a reconstruction of the only known baby *Triceratops*, the same one that kicked off the age series at the Museum of the Rockies. I wouldn't say that baby *Triceratops* were ugly. Not exactly. The little three-horned dinosaurs had an unconventional charm— the sort of so-homely-it's-kind-of-adorable look shared by manatees and shar-peis. Goodwin, the museum's assistant director, keeps the original in a file drawer downstairs.

Goodwin's office is a typical paleontologist's workspace. The dull-white room, washed with the irritating glare of fluorescent bulbs, is festooned with journal reprints, specimen casts, books, and fossil specimens. I try to suppress my fanboy enthusiasm as

Goodwin picks up part of the disarticulated *Triceratops* skull he happens to have lying on his desk. The piece is one of the brow horns, but it's not quite like the decorations on the skull in the library. These horns are curved slightly backward—a feature Goodwin and Horner had determined was an indicator of adolescence.

Goodwin explains that no two *Triceratops* skulls are exactly alike. If all the *Triceratops* skulls were assembled in one place, for example, we'd be able to see that the shapes of the frills and horn curvatures differed from one animal to another. This variation tricked paleontologists for many years, as evidenced by the fact that different authorities named *twelve* different *Triceratops* species. During the 1980s and 1990s, however, paleontologists figured out that many of these species were invalid. Most of the anatomical differences were due to individual variations and changes in the course of growth. Ultimately, the list was narrowed down to the only two *Triceratops* species we recognize today—Marsh's original *Triceratops horridus* and *Triceratops prorsus*. *Triceratops* changed dramatically as they grew, and the infant skull in Mark Goodwin's office underscores to me just how mind-bogglingly dramatic the transformation must have been.

Goodwin walks around his desk to a white set of metallic drawers pushed against the wall behind me. When he gently pulls out the appropriate shelf, I'm not entirely sure what I'm seeing. Set in boxes on a bed of foam are various tan-colored bones, many of which have obviously been carefully reassembled from a cracked and broken specimen. But three large pieces right in the middle offer a clue to what they are. The long, slightly curved bones on the left and right were the sides of a tiny frill, and the center portion made up the ornament's midline. The isolated pieces are the original parts of the reconstructed *Triceratops* skull I had seen upstairs.

I feel a little embarrassed that I didn't immediately recognize the baby *Triceratops* sitting right in front of me, but I wasn't the only one to miss the connection. Goodwin explains how the bones were originally thought to belong to a pachycephalosaur—one of the thick-skulled bipedal dinosaurs that roamed North America at around the same time. It wasn't until 2006 that Goodwin and his colleagues recognized the pieces for what they really were. I want to pick up a parietal—a bone decorated with tiny, arrow-shaped ornaments that makes up the middle part of the frill—but I try to hold myself back from asking to do so. I've had nightmares about dropping important specimens, so maybe it's best to let resting dinosaurs lie.

Goodwin takes me back up to the library before I can drum up the courage to ask. There's something about the authentic *Triceratops* skull on display that he wants me to see. Despite its size and seemingly mature anatomy, the bigger *Triceratops* skull still shows signs of change. The clues are in the frill. While the dinosaur's frill is still solid, Goodwin points out, there are two indentations on either side of the midline ridge. If Scannella and Horner are right, these were areas where the frill was thinning to create the holes seen in the skull of *Torosaurus*. The bone here was rapidly being resorbed and reshaped to create a very different look.

Such a late-life transformation is awfully strange. If dinosaurs started reproducing early, as research by Berkeley's own graduate students Sarah Werning and Andrew Lee has shown, then why would they develop flashy "Hey, look at me!" ornamentation at such a late age? Wouldn't such characteristics—perhaps useful for indicating sexual maturity and dominance—develop earlier? I ask Goodwin if any modern animals show such late-life ornamental change. "Hornbills and cassowaries," he replies, referring to birds with prominent crests on their heads. Perhaps, as in these birds, the ornaments of *Triceratops* continued to change throughout the animals' life.

Indeed, a different dinosaur at another museum might throw additional support to Scannella and Horner's hypothesis that *Triceratops* was always changing. Inside a glass case at the Smithsonian's National Museum of Natural History rests the enigmatic skull of a dinosaur named *Nedoceratops*. The skull looks more or less like a *Triceratops*, but with a few important differences. It lacks a nasal horn and has several holes in its frill, including a parenthesis-shaped fenestra in one of its parietal bones. Paleontologists have often treated these features as pathologies or unique characteristics that distinguish this dinosaur from its contemporaries, but according to Scannella and Horner, some of the features illustrate a major skull revamp. The hole in the parietal bone, for example, could just be a transitional stage between the solid frill of younger *Triceratops* and the gaps in the frill of older forms.

If Scannella and Horner are correct, the *Triceratops* at Berkeley and the *Nedoceratops* at the Smithsonian are the critical specimens that show how the classic *Triceratops* skull transformed into what we've previously called *Torosaurus*. In their 2010 paper, the paleontologists concluded that these dinosaurs didn't just change between birth and the time they started reproducing, but kept changing almost to the end of their lives.

Leave it to sensational news items to misconstrue the findings. Even though Scannella and Horner affirmed that *Torosaurus* and *Nedoceratops* would be sunk, and specimens given those names would now go by *Triceratops*, many bewildered journalists reached the opposite conclusion. The dinosaur name game distracted from the major conclusion of Scannella and Horner's study, and the confusion stemmed from the revised growth series. Since *Torosaurus* represented the fully mature form of the dinosaur, with the biggest *Triceratops* specimens being younger animals, some journalists assumed that *Triceratops* had been erased, and the name of the mature form was the right one. "The *Triceratops* Never Existed," mourned Gizmodo, as did various other sources, from CBS News to the *San Francisco Chronicle*.

The supposed demise of the iconic dinosaur even appeared as a riddle on NPR's weekly quiz show *Wait Wait . . . Don't Tell Me!* ("My third horn will go—yes, I swear it drops, the other two are a shy pair of flops. This dino's not scary, I'm just ordinary, there's no such thing as a . . ." "*Triceratops*," Carl Kasell teased.) *Triceratops* fans were outraged, of course. A "Save the *Triceratops*" Facebook group bubbled up in pointless protest, and comment threads on dozens of websites roiled with the complaints of amateur dinosaur experts who weren't about to let scientists take away their child-hood memories. My favorite objection was a T-shirt design that depicted *Triceratops* alongside Pluto, the former planet. *Triceratops* is a classic dinosaur, they screamed, and it's one of the first we encounter as children. To many who cherish dinosaurs, this felt like "*Brontosaurus*" all over again.

The concerned dinosaur fans eventually understood that *Triceratops* wasn't going anywhere. As Scannella and Horner's study pointed out, *Triceratops* was named first, and so it had sci-entific priority over *Torosaurus*. Only the fans of *Torosaurus* (like myself) had reason to worry. Almost no one gave a damn about *Nedoceratops*.

Of course, just publishing a new idea doesn't mean that scien-tists must agree with the proposal. Scannella and Horner's paper was not a statement of fact, but one hypothesis based on the cur-rent spread of data. Offering a contrary view, ceratopsian expert Andrew Farke pointed out that no one has ever documented the kinds of changes required to transform a solid-frilled *Triceratops* skull to a *Torosaurus* skull before. The enigmatic *Nedoceratops* aside, paleontologists haven't uncovered the critical skulls that show how *Triceratops* could have morphed into *Torosaurus*. What's more, Farke has noted that a *Torosaurus* specimen at Yale's Peabody Museum is a candidate for a young individual of this animal. If the skull really does belong to a juvenile or sub-adult *Torosaurus*, then that means the dinosaur *must* have been a distinct, rare spe-cies that lived alongside *Triceratops*.

The debate over *Triceratops* is just warming up, and is just part of a bigger push to document how dinosaurs grew. It stems from the ongoing point of contention in dinosaur biology that for decades tangled up the unrecognized stages of dinosaurs' growth and aging with the discovery of new dinosaur species. As we have started to realize how radically dinosaurs could change over their life cycles, dinosaur taxonomists have had to backpedal, folding many presumed species into others of which they turn out to be only older or younger forms.

While *"Brontosaurus"* was unquestionably the most famous dinosaur to vanish, many others have disappeared, too. In 1975, for example, when Peter Dodson surveyed the diversity of crested hadrosaurs, he found that *"Procheneosaurus"* was just a juvenile of another hadrosaur species. The characteristics that appeared in adults—particularly the ornate crests—were relatively undeveloped in the juvenile form. The anatomical differences were due to growth, he realized, not membership in different species. The same was true of *"Brachyceratops,"* which are really juveniles of other horned dinosaurs, such as *Rubeosaurus*, that grew to much larger sizes. By reexamining the anatomy of many small dinosaurs, paleontologists have discovered that what we once thought were diminutive species were actually the youngsters of even bigger dinosaurs.

In recent years, paleontologists have turned to additional lines of evidence—inside dinosaur bones—to investigate how dinosaurs aged. Researchers can examine whether certain bones were fused or not to estimate a specimen's stage of life, and dark bands inside many fossil bones can help paleontologists determine how old dinosaurs were when they died. These prominent rings formed when dinosaurs slowed their growth—most often during harsh winters or dry seasons when food was scarce—and this pattern allows paleontologists to roughly measure a dinosaur's age. Likewise, by looking at the kind of tissue between the bands, paleontologists can detect whether a dinosaur was still rapidly growing when it perished, or if its growth had slowed down due to maturity. The microscopic structure of dinosaur bone can tell us much about how these animals grew, and, by using these clues, paleontologists have found that *Triceratops* wasn't the only dinosaur to undergo rapid, sweeping transformations.

A few years before the *Triceratops* debate broke out, Goodwin and Horner proposed that two spiky, dome-headed dinosaurs—*Dracorex* and *Stygimoloch*—were really just younger versions of *Pachycephalosaurus*. All three lived in the same place at the same time, and evidence indicates that the cranial spikes of the younger ones were resorbed as the animals matured and a thick dome formed over the skull. *Voilà!*—three different dinosaurs filled out the life history of just one form. Similarly, the shovel-beaked hadrosaur *Edmontosaurus* is extremely common in the Late Cretaceous strata of Montana and the surrounding region, and we thought that this dinosaur lived alongside an even larger hadrosaur appropriately named *Anatotitan*. But according to Nicolás Campione and David Evans, *Anatotitan* was actually the bigger, more mature form of *Edmontosaurus*.

Even the tyrant king himself figures into the swelling argument over how dinosaurs grew. In 1946, Charles Gilmore described a strange tyrannosaur head found in Montana. The skull was small, with a long and low profile, and Gilmore thought that he had found a new species of the dinosaur genus *Gorgosaurus*—a tyrannosaur that was a bit slimmer and more agile than *T. rex*. He named the animal *Gorgosaurus lancensis*. Four decades later, however, paleontologists proposed that the skull actually represented a "pygmy tyrant" that sprinted through the same haunts as its more famous relative. On the basis of the skull Gilmore collected, they rechristened the animal *Nanotyrannus*, and that choice has opened up a debate about how drastically tyrannosaurs changed as they grew.

As with *Triceratops*, paleontologists regularly mistook young tyrannosaurs—leggy, gangly dinosaurs with shallow skulls that

Young dinosaurs didn't look exactly like their parents. This reconstruction of a juvenile *Tyrannosaurus rex,* **on display at the Natural History Museum of Los Angeles, shows how much the little tyrants changed as they aged.** (Photograph by the author)

lacked the crushing capabilities of the adults—for distinct genera or species. In addition to *Nanotyrannus,* paleontologists named at least four different tiny tyrannosaurs, including the recently announced *Raptorex,* on the basis of small, often incomplete, specimens. All of these have been discarded. As Thomas Carr and other experts have discovered, the younger growth stages of some of the last tyrannosaur species often had long, low skulls that vaguely resembled earlier forms of tyrannosaur. As the animals grew, however, their skulls deepened, and the number of teeth in their mouths changed as their dental equipment transformed from slashing blades into crushing railroad spikes. "Some people become lobotomized around tyrannosaurs" and confuse characteristics of young tyrants for signs of new species, Carr told me, and our unfettered love for them has led us to overlook differences due to growth or variation so that we can crown a new tyrant species. As it stands now, there is only one tyrant dinosaur known from the latest Cretaceous of North America, and that is still our old faithful *Tyrannosaurus rex.*

I've always found it intriguing that the most heavily debated cases of metamorphosis all concern Hell Creek Formation dinosaurs. *Triceratops, Edmontosaurus, Pachycephalosaurus,* and *Tyrannosaurus* were all among the last non-avian dinosaurs. This affects the debate over just how catastrophic the end-Cretaceous mass extinction really was. If *Triceratops, Torosaurus, Dracorex, Stygimoloch, Pachycephalosaurus, Edmontosaurus, Anatotitan, Nanotyrannus,* and *Tyrannosaurus* all coexisted, then dinosaurs maintained their diversity right up until the very end. But if all the proposed growth series are correct, this diversity is instantly cut in half, and the Late Cretaceous looks emptier.

North America's Late Cretaceous dinosaurs have been the most intensely scrutinized so far, but the argument affects all dinosaurs. Now that we have a better idea of how much dinosaurs changed, we are better equipped to spot infant dinosaurs in the field and in museum collections, too. Our newfound interest in nesting and growth isn't just about mapping the trajectory of dinosaur lives. The narrative of how dinosaurs started life and how they changed after hatching incorporates various threads of dinosaur biology—from their social behavior to their physiology. In fact, the way dinosaurs grew up also has important implications for one of the most puzzling aspects of paleobiology: How did dinosaurs like *Apatosaurus* get so big?

Five

Jurassic Thunder

From the time I scampered through the American Museum of Natural History's fossil halls as a child, I wanted to find a way to visit the institution's dinosaur warehouse. The mounted skeletons were great—some of the finest dinosaurs ever uncovered—but I desired more. I wanted to see the scores of dinosaurs that had never made it to the exhibit galleries. And the bigger the bones, the better.

Public displays are where dinosaurs live. Reconstructed in bone, fiberglass, and plaster on painstakingly crafted metal armatures, the Mesozoic celebrities pose for their adoring public. But no museum has space to display everything it has obtained, or even the most fascinating fossils in its care. If museums put their entire fossil collections on exhibit, visitors would have to slog past rows upon rows of mammal teeth, turtle shells, and petrified bone fragments before reaching the relatively few dinosaurs complete enough to even consider mounting. Not to mention the fact that it's hard to study the intricate details of a dinosaur's anatomy when the skeleton is locked into a mounted pose. If we're going to extract the secrets of dinosaur lives, paleontologists need easy access to the ever-increasing wealth of bones in safely kept museum back rooms. Display halls present the products of paleontological research, but the long rows of metal shelves heaped with fossils and drawers cradling delicate specimens provide the basis for much of what we have learned about prehistoric life.

My dream finally came true on a crisp March evening in 2011. The AMNH was preparing to debut its World's Largest Dinosaurs exhibition, and invited bloggers and Twitter addicts like myself to preview the show. The C train from Penn Station to the museum's uptown stop couldn't go fast enough. In addition to a sneak preview of the displays, the program promised a tour of the AMNH's private fossil storerooms. After two decades of waiting, I'd finally get my chance to stroll, slack-jawed, through the institution's cherished dinosaur storehouse.

When I reached the meet-up in the fourth-floor Hall of Saurischian Dinosaurs, I wasted no time signing up for the earliest collections tour on the schedule. I was sure the *Gorgosaurus* and *Apatosaurus* on display would understand—they were old friends from long weekend afternoons in the museum's galleries, and this was my chance to see their disassembled counterparts. Senior scientific assistant Carl Mehling met my tour group after a quick elevator ride down to the museum's extensive basement. Carl was an amiable host, even when I made him a little nervous by leaning over a beautiful articulated skeleton of an oviraptorid, nesting chicken-style over a brood of eggs, to get a few close-ups of the dinosaur's wishbone. The skeleton, part of a pair, was a gorgeous specimen from the Cretaceous of Mongolia. The dinosaur's cream-colored bones popped out against the rusty orange of the surrounding rock.

But most of the storeroom was home to sauropod bones. Individually cataloged and nestled on beds of foam, long strings of vertebrae were arranged along the shelves, as if the museum had been playing a game to see how many sauropods they could fit into a single room. This was paleontological Tetris. Each and every bone was a beautiful monument to evolution. On one shelf sat a single neck vertebra of *Barosaurus*—a particularly slender giant that reached over eighty feet in length. Compared to the short stack of bones that make up my own neck, the single dinosaur bone is a natural masterwork, beautifully sculpted into an elongated central shaft indented by air pockets and ornamented with short wings of bone.

I even bumped into an old friend as I poked around the lower shelves. Tucked away among miscellaneous bones and casts was a copy of the museum's outdated *"Brontosaurus"* head—a double of the tiny, grinning skull that seemed to be anchored so impossibly high above my head so many years ago. A worn, dusty card kept with the model confirmed that this was the skull of the "great amphibious dinosaur" modeled by the museum's preparation expert Adam Hermann a century before. Even though she was kicked off exhibit, I was glad to see the old girl again.

But there was another piece of dinosaur history I desperately wanted to see. Paleontologists have discovered many sauropods, but the AMNH was supposed to be the home of the biggest of the big, the ultimate in superlative dinosaur size. If I thought he could have helped, I would have asked Mehling if he could locate the fossil, but the bone I was looking for went missing right around the time that Hermann created that *"Brontosaurus"* skull. I was after *Amphicoelias fragillimus*—what may have been the largest creature ever to walk the Earth.

If *Amphicoelias* was hiding among the crowded and dusty shelves, it wouldn't have looked like much. Just a strange lump of bone, albeit one four and a half feet high. The long-lost fossil was a yet another Bone Wars dinosaur—much like *Apatosaurus* and *Triceratops*—but this creature was one of Edward Drinker Cope's prizes. Like many dinosaurs first discovered in the nineteenth century, though, *Amphicoelias* has a complicated history.

In 1878, when his competition with O. C. Marsh was at full intensity, Cope announced the discovery of what he believed to be an absolutely enormous dinosaur. He detailed the find in a four-page paper, "On *Amphicoelias*, a Genus of Saurian from the Dakota Epoch of Colorado," and that brief missive is a perfect example of how quickly the dueling paleontologists created new dinosaurian identities on paltry remains. On the basis of just a few bones—including a femur, several tail vertebrae, and part of a hip—Cope thought he could distinguish two different species of *Amphicoelias*, and in a move sure to frustrate dyslexics, named them *A. altus* and *A. latus*. Both species were similar to previously discovered sauropods such as *Apatosaurus*, but they were considerably larger. The femur Cope assigned to *A. altus* was six feet, four inches long. This dinosaur was surely a giant among giants.

Another *Amphicoelias* specimen was bigger still. In a dispatch published in August that same year, Cope wrote, "I have recently received from my indefatigable friend, Mr. O. W. Lucas, the almost entire neural arch of the vertebra of the largest saurian I have yet seen." This was just a scrap—the uppermost portion of a single vertebra, once connected to the rounded centrum—but it was a *big* scrap. Cope believed this bone belonged in the middle part of the spinal column of a new species of *Amphicoelias—A. fragillimus*. When complete, Cope ventured, the bone might have been at least six feet tall, and a back-of-the-envelope calculation based upon the skeletons of other dinosaurs led Cope to believe that the animal's femur would have been over twelve feet high. No dinosaur was bigger, and, in a bit of extra speculation, Cope cited the rough resemblance of the partial vertebra to the backbones of deep-sea fish as evidence that "these beasts may have walked in deep water and browsed on precipitous shores." Though he based it on frustratingly fragmentary evidence, Cope could now claim the largest dinosaur in his battle with Marsh. Even the giants *Diplodocus* and *"Brontosaurus"*—rapidly described by Marsh within the following year—did not come close to the dimensions of *Amphicoelias fragillimus*.

If Cope was correct, in fact, *Amphicoelias fragillimus* was the

largest of all dinosaurs. (The fossils from *A. altus* and *A. latus* turned
out to be *Diplodocus* bones, and so paleontologists recognize only
A. fragillimus today.) Based upon the proportions of dinosaurs such
as *Apatosaurus* and *Barosaurus*, the paleontologist Kenneth Carpenter estimated that *A. fragillimus* would have stretched an impressive 190 feet long—nearly twice as long as the next contender for
prehistory's longest dinosaur. If you stood at the tip of the dinosaur's nose and set out at a leisurely walking pace, it would take
over half a minute to reach the tip of its whiplash tail.

The only evidence of the dinosaur's existence disappeared decades ago. After Cope's death in 1897, many of his fossils were
acquired by his friend and pupil Henry Fairfield Osborn at the
American Museum of Natural History. When Osborn and colleague Charles Mook sifted through Cope's collection in 1921 in
an effort to reanalyze some of the older finds, though, *Amphicoelias*
was gone. Maybe the bone is lost among the shelves somewhere,
perhaps someone destroyed the fossil, or the bone may have simply crumbled into scores of irreparable pieces in the days when
preservation techniques were still crude. No one knows. And in
130 years of prospecting among the West's Jurassic badlands, no
fossil hunter has ever found another specimen. Even the original
quarry is tapped out. When Carpenter tried to relocate the *Amphicoelias* site in 1994, there was no sign of the dinosaur anywhere.
Most of the skeleton may have even eroded away by the time Lucas discovered the bonebed in 1877—the neural arch was all that
was left of a true titan. How could such a gigantic dinosaur have
so thoroughly disappeared?

The same question could be asked of other giants. Our frustratingly incomplete inventory of the sauropods makes the question of
who was the largest dinosaur of all time exceedingly difficult to
resolve. No surprise, then, that a series of giant dinosaurs have tried
to lay claim to the title of the biggest of the big.

Before I knew about *Amphicoelias*—the dinosaur that got away—
I was told that the eighty-five-foot-long, twenty-ton *Brachiosaurus*

was as big as dinosaurs got. The dinosaur was so huge, a Time Life Young Readers Nature Library book told me, that *Brachiosaurus* must have been bound to deep Jurassic lakes. Only later would I learn that *Brachiosaurus* and kin were bound to the land. The air sacs that pervaded their skeletons made them surprisingly buoyant for their size, and even if they had somehow managed to stand still up to their noses in water, the pressure would have fatally constricted their chests. Improbable though it seems, sauropods moved their bulk about on dry land.

Not long after I learned about *Brachiosaurus* from outdated books, I heard about three recently discovered sauropods that have robbed the "arm lizard" of the Largest Dinosaur title. Two of them stand at the Museum of Ancient Life in Lehi, Utah. A twenty-minute drive south of Salt Lake City, the museum boasts an enormous collection of casts placed in eerie reconstructed environments where skeletal crocodiles tear at a fallen *Stegosaurus* and a pair of *Tyrannosaurus* snarl at each other over a kill. The skeletons don't simply pose—they seem ready to step off their pedestals and start devouring guests as if they were in a direct-to-DVD horror flick. Not all the exhibits feature Mesozoic death, though. Other dinosaurs simply posture for their admiring public, and the largest of all are a pair of sauropods in the first dinosaur hall. Looming over an archway leading to the next exhibit hall is the dinosaur formerly known as *Ultrasaurus*. The Brigham Young University paleontologist Jim Jensen dug up a few parts of this dinosaur from Dry Mesa, Colorado, in the late 1970s. The sauropod was heralded as the largest ever before it was even properly described. When I tuned in to the Walter Cronkite–hosted documentary *Dinosaur!* in third grade, the show used some low-grade special effects to manifest the dinosaur at BYU's football stadium, with fleeing cheerleaders for scale. Looking at the dinosaur's fossils in the museum, there's no doubt that it was enormous. Below the facsimile mount is the reconstructed shoulder blade that inspired Jensen's hype about the dinosaur; the lone bone, riddled with

A reconstruction of *Diplodocus carnegii*. This 80-foot dinosaur was among the largest animals of all time, although there were a handful of giants that got even bigger. (Illustration by Scott Hartman)

cracks and breaks, was just as tall as Jensen was. And up against the far wall is another of Jensen's discoveries from Dry Mesa: a 100-foot-long skeleton of *Supersaurus*. The huge dinosaur, similar to an overgrown *Diplodocus*, runs almost the entire length of the exhibit space.

The paleontologist David Gillette added another contender in 1991—a slender sauropod said to extend 170 feet from nose to tail tip. This was *Seismosaurus*. I loved saying the name of this dinosaur and its gargantuan competitors. *Ultrasaurus*, *Supersaurus*, and *Seismosaurus*—the names alone deeply resonate when you say them, a portent of a giant's arrival. But only *Supersaurus* remains today. These three dinosaurs—named from scrappy portions of skeletons—have a complicated and intertwined taxonomic history that is best left to technical journals. Ultimately, three dinosaurs entered, and only one left. That big *"Ultrasaurus"* scapula came from a *Brachiosaurus* that was not as gigantic as Jensen had supposed, and Gillette's *Seismosaurus* turned out to be a big *Diplodocus* that topped out at about 110 feet, given a substantial boost by the very long tail typical of the genus. Only *Supersaurus* has survived the revisions and nomenclatural shifts.

The problem is that we don't know exactly how big the largest dinosaurs grew to be. *Supersaurus* and Gillette's big *Diplodocus* are top contenders, but as yet no one knows how large these dinosaurs really were. Even the dinosaur that paleontologists most often cite as the biggest of all—*Argentinosaurus*, thought to have been about 100 to 120 feet long—is incompletely known. Paleontologists can only estimate the size of the most gargantuan dinosaurs based upon what we know about their better-known relatives. The largest dinosaurs have been named on some of the most fragmentary material. This dearth of complete skeletons makes perfect sense. Small, delicate dinosaurs can quickly and easily be covered in sediment. A little bit of saturated mud or wet sand is all that's needed. But a 100-foot-long, 60-plus-ton behemoth is another matter. A local flood or other natural disaster capable of moving massive quantities of sediment would be needed to entomb such an animal, and that's to say nothing of the scavengers which undoubtedly attended the burial of the biggest dinosaurs. (The death of one enormous dinosaur enriched the lives of many carnivores.) There were many more giant sauropods than ever found their way into the fossil record, and even among those left behind as fossils, few died in circumstances amenable to high-quality preservation. If a nearly complete skeleton of *Supersaurus*, *Argentinosaurus*, or even *Amphicoelias* exists, no one has found it yet.

When considered as a whole, all of the largest dinosaurs seem to be around 100 feet long, and maybe a little more. (Only *Amphicoelias* is estimated to be considerably bigger.) If so many of the largest dinosaurs topped out at around the same size, this might represent the upper boundary of just how large a dinosaur could get before collapsing upon itself.

And not all sauropods were big. Indeed, most species never approached that 100-foot upper bound, and on prehistoric islands, some stranded populations of sauropods even became dwarfs that reached only about 20 feet in length. What makes a sauropod a sauropod isn't size, but anatomy. And these dinosaurs were *weird*. As paleontologists discover more, the dinosaur lineages Cope, Marsh, and other researchers found early on seem rather plain, but I still believe that those sauropods were some of the most bizarre dinosaurs of all.

Consider *Apatosaurus*. This hefty herbivore, familiar to just about everyone who's ever heard the word "dinosaur," might seem rather vanilla. Lacking spikes, horns, armor plates, a crest, or any other bizarre ornamentation, the elegant sauropod might not strike us as nearly as strange as an armor-encased ankylosaur, or as unusual as a feather-covered, sickle-clawed deinonychosaur. But their familiarity belies how little we know about the animal's natural history. Sauropods are so marvelous not just because of their size, but because they are improbable creatures in so many ways. Almost everything about sauropod biology is a puzzle, especially their most distinctive feature.

The necks of sauropods are ludicrous monuments to evolution. Any giraffe would be green with envy at the ability of these dinosaurs to pluck fodder from high in the trees or sweep their heads over fern-covered savannas to suck up mouthful after mouthful of vegetation. Their necks also beautifully demonstrate the jury-rigged nature of evolution while simultaneously refuting the notion that some divine Artificer intelligently designed organic life. As Matt Wedel remarked in a recent paper on these dinosaurs, part of their necks were a fantastic "monument of inefficiency."

Evolution does not operate in the best of all possible worlds. As the paleontologist Stephen Jay Gould remarked in his famous essay on the panda's "thumb"—a modified wrist bone used by the black-and-white bear to grip bamboo—"Odd arrangements and funny solutions are the proof of evolution—paths that a sensible God would never tread but that a natural process, constrained by history, follows perforce." The recurrent laryngeal nerve of sauropods is another beautifully complicated example of circuitous anatomical solutions. The unwieldy nerve wasn't unique to huge dinosaurs, but a trait they shared with all other four-limbed vertebrates, inherited from a common ancestor.

Over 375 million years ago, creatures like the flattened "fisha-pod" *Tiktaalik* formed the basis for what would eventually become a fantastic radiation of animals called tetrapods. From early amphibians to dinosaurs to vertebrates that secondarily lost their limbs—snakes and whales, for example—tetrapods are a widely varied group of animals. And all tetrapods share the recurrent laryngeal nerve. This string of neurons—which runs from the brain down along the neck, and allows both sensation and motor response by the larynx—started off as a close connection between the head and heart among the early four-limbed fish that dragged themselves through the Devonian mud. As tetrapods flourished and forms with longer necks evolved, however, the nerve had to stretch to keep brain and throat in contact—and a twist in the nerve's pathway made the structure twice as long as it had to be. The nerve takes an extended path from the brain into the chest cavity, where it loops around a series of major blood vessels called aortic arches and winds back up the length of the neck, a course resembling a drawn-out U. This wasn't an issue for creatures with compact bodies, with brains not so very far from their hearts, but it becomes ridiculous in extremely long-necked tetrapods. A giraffe with an eight-foot-long neck carries a laryngeal nerve about sixteen feet long. This boneheaded "design" is a hallmark of evolution—what already existed is modified and co-opted into new forms.

Sauropods had the most stupendously inefficient nerve paths of all. The longer the neck, the further the nerve has to stretch, and sauropods unquestionably had the longest necks of any creatures. According to Wedel's calculations, an adult *Supersaurus* is estimated to have had a neck about 46 feet long and would therefore have had a recurrent laryngeal nerve almost as long as its body at 92 feet long. "The existence of 28 m neurons in the RLN of *Supersaurus* may seem fantastic," Wedel writes, "but they appear unavoidable given what we know of tetrapod embryology and evolution."

Wedel also points out that sauropods most likely had even longer nerve cells. Just think about the nerves at the end of a sauropod's tail that would have had to convey signals triggered by touch all the way back to the dinosaur's brainstem. Wedel notes that such a connection suggests that that the largest of modern-day whales have neurons that reach over 90 feet, and slightly larger sauropods would have required even longer ones. No one has actually seen these cells, but, as Wedel hypothesizes, such fantastically long neurons should exist. The neurons in a dinosaur like *Supersaurus* could have been the longest cells of all time, creating quite a problem for these dinosaurs. If *Amphicoelias* was a real dinosaur and truly reached over 160 feet, Wedel points out, then even nerve impulses traveling at over 330 feet per second would have required several tenths of a second to reach the brain. While this doesn't mean that sauropods were so slow that predatory dinosaurs could nibble on the tails of the giants without the behemoths knowing, there definitely would have been a delayed response to signals from the outside world. As Wedel notes, this might be a hint that the largest dinosaurs were reaching the upper limit of how big it was possible to get.

And for their size, sauropods had ridiculously small heads. I have an *Apatosaurus* skull cast in my living room. Even though my wife specifically told me, "Don't bring home any dinosaurs," when I went to the estate sale of the former state paleontologist of Utah, James Madsen, Jr., I couldn't resist the full-size replica of an *Apatosaurus* skull. The carbon-copy cranium belonged to an animal more than 80 feet long, yet I was able to comfortably cradle the dinosaur's head in my arms as I happily walked it out to my car. This is a tiny head for such a big animal. And, even worse, this was not a skull suited to chewing. Much like its contemporary cousin *Diplodocus* and other sauropods, *Apatosaurus* had only a short row of pencil-shaped teeth at the front of its square muzzle. How did *Apatosaurus* consume enough food to fuel itself? I'm an unabashed fan of evolution, but sometimes I wish nature's mysteries weren't so damn hard to solve.

Our mammalian bias often gets in the way of understanding dinosaurs. We're chewers, and we expect dinosaurs to have done the same. But that wasn't the case at all. Dinosaur jaws plucked, sliced, cleaved, ripped, and otherwise cropped food, but then they immediately horfed their meals down. Sauropods must have been champs at this. *Apatosaurus* didn't stand on Jurassic floodplains grinding down ferns and conifer branches like some sort of Jurassic cow. The dinosaur's bad table manners allowed the sauropod to suck up the vast amounts of succulent green food it required to survive. Exactly how much food *Apatosaurus* and similarly sized dinosaurs needed is a matter of physiology, though, and that is one of the most frustrating of all dinosaur mysteries.

When I first met dinosaurs, the Dinosaur Renaissance was still in the process of revitalizing them. Plodding Mesozoic idiots on television and in my library books were gradually being replaced by colorful, clever dinosaurs that looked limber enough to turn cartwheels over the ancient landscape. In the pop-sci parlance of documentaries, dinosaurs had gone from being typical cold-blooded reptiles to hot-blooded creatures unlike anything before or since. But while paleontologists generally agreed with the new image of dinosaurs as far more complex than anyone had previously understood, they fought like hell over the particulars of dinosaurian biology.

Unless you're describing a romance novel, "hot-blooded" is an awful term. It doesn't tell you anything about an animal's biology. With a more or less constant body temperature of about 98.6 degrees Fahrenheit, I'm unquestionably hot-blooded, but a lizard that lies out in the sun for long enough will also warm to an active, hot-blooded state. An animal's physiological profile isn't equal to its body temperature. The distinction lies in the varied biological mechanisms involved in how that body temperature is maintained. For dinosaurs, this means we have to figure out whether they regulated their body temperatures internally, whether they maintained constant body temperatures, and whether they had high or low metabolic rates. Those three features create an outline of a creature's physiology, and paleontologists have struggled to understand these details of dinosaur biology.

Lacking a time machine and a thermometer, we can't measure

dinosaurian temperature directly. That may be for the best. The paleontologist Edwin Colbert and his collaborators had enough trouble taking the temperature of American alligators in an experiment meant to sketch what kind of physiological strategy dinosaurs employed. Three decades before the dinosaur temperature debate exploded, Colbert and colleagues traveled to Florida to measure the body temperatures—via the cloaca—of various small American alligators as the scientists moved the archosaurs between sun and shade. Since alligators are ectotherms, and therefore rely on their environment to regulate their body temperature, the scientists wanted to see just how fast the crocodylians warmed up and cooled down as a way to create a model for how much time even larger dinosaurs might have spent sunning themselves. The researchers even went as far as to create little wooden armatures—which looked like torture devices—that they used to manipulate little alligators into dinosaur-like positions to see if posture made any difference. The illustration of the boxes included in the paper looks like a snapshot from an alligator crucifixion.

The experiment didn't go exactly as planned. Two alligators died from prolonged exposure to the sun. Even supposedly sun-loving archosaurs could fatally overheat. And, not surprisingly, warm-up and cool-down time varied based on the alligator's size. The smaller alligators warmed up and cooled down faster than their larger counterparts, thanks to their smaller volume. (It's the same concept that explains why a small plate of leftovers heats up fast in the oven but a plump Thanksgiving turkey takes hours.) The results didn't solve any dinosaurian mystery. When Colbert applied the patterns to dinosaur body sizes, *Apatosaurus*-class sauropods would have required a whole day in the sun before they fully warmed. And the same principle held for cooling down. If an *Apatosaurus* started to overheat, the dinosaur would have had an awful time trying to dump the excess heat and might have died, just like the small alligators. An ectothermic lifestyle reliant on sunbathing just didn't work for big dinosaurs, and small dinosaurs were so lightly built and agile that they didn't seem to need any warm-up time. Even as paleontologists, artists, and animators still imagined that sauropods required warm ponds to survive, evidence was mounting that dinosaurs needed a fresh look.

Dinosaur imagery changed before paleontologists really understood how sauropods and their various relatives functioned. The paleontologist Bob Bakker, in particular, pointed out that dinosaurs grew fast, had upright limb postures consistent with an active lifestyle, and had population structures more reminiscent of mammals than reptiles—all clues that dinosaurs may have kept their internal fires burning at high, constant temperatures. All those lines of argument still hold true. And that's not all. Many dinosaurs were covered in insulating feathers, and entire communities of dinosaurs flourished in polar habitats, where they undoubtedly faced long, snowy winter nights. The more we learn, the more it's apparent that dinosaurs had the internal mechanisms to run hot. Dinosaurs were definitely not sluggish ectotherms restricted to a perpetual muggy Mesozoic summer, as brought to life in Disney's *Fantasia* and Rudolph Zallinger's *Age of Reptiles* mural at Yale's Peabody Museum of Natural History.

Mammals have even helped researchers investigate the physiology of dinosaurs. In 2012, a landmark study of mammal bones showed that ruminants—hoofed herbivores with an even number of toes on each foot—had lines of arrested growth in their bones. These bands are signs of a seasonal slowdown, and naturalists had previously thought that they were present only in ectothermic organisms whose physiology fluctuated with the surrounding environment, such as crocodiles. Since dinosaurs had these same lines, some paleontologists had suggested that dinosaurs were more like reptiles than like mammals or birds, but the new study struck down this line of argument. As the paleontologist Kevin Padian remarked in an opinion piece on the research, the study showed that dinosaurs "were anything but typical reptiles." Now we know that at least some mammals, too, grow rapidly when times are good and slow their growth during dry or cold seasons, when resources are scarce.

Together, these lines of evidence strongly suggest that dinosaurs were active, fast-growing creatures that would have required high metabolic rates. As the paleontologist Stephen Brusatte wrote after reviewing the evidence accumulated so far, "What seems clear . . . is that dinosaur physiology and metabolism was more similar to that of living birds and mammals than living reptiles." Much of dinosaur physiology remains the subject of fierce debate, but with a tenure spanning the past 230 million years, dinosaurs must have had a physiology that was immensely adaptable.

Of course, there was no single physiological profile that fit all dinosaurs. Dinosaurs were so diverse and disparate—just think of the differences in size and shape between *Supersaurus*, the tiny fluffball *Sinosauropteryx*, and the heavily armored *Kentrosaurus*, to pick just three of hundreds of genera—that lineages varied just as dramatically as living mammals do (say, a bat, a dolphin, and an elephant). And sauropods are among the most vexing of all. Little sauropods grew at such an astonishing rate that there's almost no way they could have done it without having highly active metabolisms and probably maintaining high body temperatures—but that ability could come at a cost at larger sizes.

Some paleontologists have highlighted sauropods as perfect examples of gigantothermy—a strategy in which body temperature remains more or less constant because of sheer size, rather than metabolism. A dinosaur the size of *Apatosaurus* would have had a hard time gaining or losing heat, but if the dinosaur was ectothermic and conserved heat rather than actively generating it, the giant would have had a little more physiological leeway before being in danger of overheating. As a corollary to that, some researchers have cast sauropods as giant walking compost heaps—kept warm by the breakdown of all that plant material inside them. But these views still see sauropods as huge reptiles rather than as the unique creatures they were. From indentations on their bones, we know sauropods had systems of air pockets along their vertebral columns—especially their necks—similar to the air sacs branching from the respiratory systems of today's avian dinosaurs. These air-filled structures not only made the dinosaur's skeleton lighter, but may have acted as a kind of air-conditioning system, as it does in birds, allowing sauropods to cope with excess heat while on the move.

Admittedly, the physiology and biology of sauropod dinosaurs is a fast-changing area of research. This much is clear, though—whatever sauropods did, it worked. They were not an evolutionary fluke, but a group in which truly gigantic size evolved multiple times, culminating in some of the largest vertebrates of all time.

As we wonder about *how*, we face the equally daunting question of *why* some sauropods became giants. Over the years, authorities have suggested an array of different ideas: that size increase was a defense against predators; that the Mesozoic atmosphere contained more oxygen, and therefore let dinosaurs breathe more efficiently; or even that the pull of the Earth's gravity wasn't as intense in the past. None of these environmental factors stand up to scrutiny. The real secret of sauropod size is an irony: they got so big because they started off small.

Contrary to the sweet stories of doting dinosaur parents in *The Land Before Time*, and even in the less well-known *Baby: Secret of the Lost Legend*, sauropods did not have single offspring that they attentively nurtured. In reality, Littlefoot's mom would have actually had many more offspring and cared for them considerably less. From fossil eggs and nests, we know that mother *Apatosaurus*, *Supersaurus*, and *Argentinosaurus* laid clutches of a dozen or two relatively small eggs.

I didn't really understand just how small baby sauropods were until I got a chance to visit the finished version of the AMNH's *World's Largest Dinosaurs* exhibit. Huge models of dinosaur organs and a life-size sculpture of *Mamenchisaurus* dominated the show, but behind the exhibits showcasing the features of the largest dinosaurs, a small display presented a model of a sauropod nest. The eggs were about the size of grapefruit, and the exhibit's baby sauropods could have curled up in the palm of my hand. The tiny dinosaurs would have provided little more than a quick snack for a passing theropod, or, as shown by lovely fossil finds in India, constrictor snakes that slithered through sauropod nesting grounds.

By starting out small, researchers have found, sauropods were freed from the biological constraints that have limited mammal body size. These constraints explain why there aren't any *Apatosaurus*-sized mammals walking around today. Taking a cue from fossil mammal expert Björn Kurtén, in 1990 the Brown University paleontologists Christine Janis and Matthew Carrano looked at the different ways the largest dinosaurs and the heftiest land mammals reproduced. While sauropods laid sizable clutches of relatively small eggs that they tended fairly briefly, if at all, elephants, giraffes, and other large mammals gestate small numbers of offspring—typically just one—for very long periods of time. And, after birth, mammal babies continue to be an energy draw on the mothers as they require milk and attention. These features of the way mammals reproduce—long pregnancies during which much can go wrong, and infants with intense and prolonged energy needs—put limits on mammals that dinosaurs did not experience.

In 2011, the zoologists Jan Werner and Eva-Maria Griebeler reinvestigated the idea proposed by Janis and Carrano. They found that dinosaurs really did have a reproductive edge. By laying eggs and leaving fast-growing offspring early in life, dinosaurs were not energetically taxed in the same way as mammals, and thus had the biological flexibility to be adapted to sizes that may represent the limit of how large terrestrial animals can get. And this reproductive and life-history difference might account for why there were so many huge dinosaurs while there have always been few behemoth-class mammals. Dinosaurs could lay frequent clutches of eggs no matter how big they were, but big mammals might produce only one offspring every few years. This makes large mammals much less prolific and less likely to recover if a population comes under stress. Slow reproduction, which limits their numbers, is a cost of large body size when parental investment is high. Dinosaurs, by contrast, experienced no such cost for being big. High reproductive throughput and low parental investment did not *require* that they become giants, but the way sauropods kickstarted the next generation made it *possible* for absolutely enormous dinosaurs to evolve.

To grow to astonishing sizes, though, hatchling sauropods had to survive. They may not have received much help from their parents. Evidence from dinosaur nest sites and from the behavior of birds and crocodylians hints that dinosaur parents looked after their nests and provided some post-hatching care, as we've seen; but baby dinosaurs may have been on their own once they left the nest, and led separate lives from adults of the same species.

Skull shape is one clue. The skull of a juvenile *Diplodocus* found at Dinosaur National Monument has a differently shaped snout than the adult dinosaurs. While mature *Diplodocus* had squared muzzles suited to cropping low-lying plants, the juvenile form had a rounded mouth adapted to browsing; a young sauropod had to be picky about what food it ate, and it plucked specific, nutritious plants to fuel its growth spurt. The juveniles had different needs, diets, and habits than the adults. And, over the past few years, paleontologists have uncovered bonebeds containing only juvenile animals. From *Triceratops* to the sauropod *Alamosaurus*, young dinosaurs seemed to stick together in their own cohorts before later joining a breeding group or going solo. Dinosaurs may have had a multistage life history that dictated whom they associated with and when. Glimmerings of their social lives are left in the rock record.

Six

Dinosaur Society

The Cleveland-Lloyd Dinosaur Quarry is easy to miss. Driving seventy miles an hour down the lonely, parched stretch of U.S. Route 6 between Price and Moab, Utah—the first place I ever saw an iconic tumbleweed float lazily across the road—it's not uncommon for dinosaur lovers like me to blow past the brown sign marking the start of a dirt road that winds past farms and rock exposures to an arid landscape of red, gray, and green badlands. Or maybe it's my fault for being distracted by the lovely scenery. The highway leading up to the quarry turnoff—part of Utah and Colorado's scenic "Dinosaur Diamond" byway system—is flanked by barren rock faces. Little domes of sediment flanking the highway give way to large swaths of exposed geologic time that glow when the evening sun strikes them at just the right angle. The panorama doesn't whisper the truth of Deep Time—it practically screams it. How anyone in this age can believe that all of this geologic grandeur was created in a matter of days is totally beyond me. The strata, deeply set in rainbow colors, highlight the almost incompressible depth of time and could never have been laid down by a mere flood. Ages are stacked upon ages, naked in the baking western sun. Time is evident everywhere.

The first time I navigated my wimpy little red sedan past the rock formations and over the rough road, I was greeted by a locked set of battered yellow gates just a mile from the quarry entrance. Apparently the site is so remote that the Bureau of Land Management keeps the visitor center open only during weekends, except in the summer season. Too bad I didn't find that out until I got there, just one day before the official summer hours started. Cursing my stupidity, I started my three-hour drive back to Salt Lake City.

I tried again the following week, anxious to see the bonebed. This time I had company: my patient wife was along for the ride. Tracey's not as dinosaur-crazed as I am—botany is her scientific love—but she'll take any chance to get out and explore Utah's scenery. Ever since we first visited Utah in 2009 and decided we wanted to move west, every trip we take is split between appreciating the local wilderness and, of course, going out of our way to see whatever dinosaurs might be nearby. It was truly a happy stroke of luck. By moving to Utah, she could immerse herself in unfamiliar ecosystems, I'd get to visit the dinosaur haunts I had yearned to see for so long, and we'd both get a chance to live out our dreams in a beautiful landscape studded with extravagant fossil riches. And while I missed the chance to visit Cleveland-Lloyd my first summer in Utah, I knew I had to take the earliest opportunity I could get during my second year to see one of the most important dinosaur sites ever found.

Thankfully, on my second attempt I found an easier way to reach the quarry, over paved roads that connect suburban Price to the llama and cattle ranches scattered around the rural town of Elmo, just outside the park. We still had a few miles on the dirt roads, and when we rolled down our windows to let the breeze make up for our broken A/C, fine dust swirled around the car, coating everything. I followed the BLM signposts toward the quarry, constantly checking the clock to see how much time we'd have before the site closed. I didn't want to waste a second on our way to the dinosaur graveyard.

•

Strangely, my wife and I were the only visitors to a long, low building of glass and stone, set among the vacant benches of a picnic area. Even though the quarry contains the remains of dozens of dinosaurs, and is the locale where several species of sharp-toothed theropod dinosaurs were first discovered, the site feels like a secret that only attracts people who already know it's there. The sign out on the highway gives nothing but the name of the quarry, a mundane-sounding title outside of the word "dinosaur." Maybe more tourists would pull off the interstate if the sign said *"Allosaurus* Death Trap This Way!"

The Cleveland-Lloyd Dinosaur Quarry contains one of the Jurassic's most perplexing mysteries. The star of that mystery stands, jaws agape, in the quarry's visitor center—and I make a beeline for the dinosaur I have driven three hours (twice!) to see. Just beyond a glass case of jet-black skulls—highlighting some of the charismatic dinosaurs found at the site—a round enclosure fences in a large *Allosaurus.* I have heard about this reconstruction, and I immediately see that this is a mix-and-match skeleton. While the theropod's body was reconstructed based on bones found at the quarry, the skull is a cast of a specimen found hours away at Dinosaur National Monument. And, oddly, the *Allosaurus* reconstruction at Dinosaur National Monument is based on fossils from Cleveland-Lloyd! Together, both dramatic bonebeds outline the life and times of the Jurassic's most prolific predatory dinosaur.

Allosaurus has often been cast as the wimpier precursor to *Tyrannosaurus.* But this isn't really fair, or accurate. Until the 1990s, paleontologists grouped all large flesh-tearing dinosaurs into a single group—the "Carnosauria." In this system, the apex predator of Jurassic North America undoubtedly set the stage for the even bigger Cretaceous hunters. But then paleontologists recognized that the Carnosauria was really a hodgepodge of very different dinosaurs that belonged to various hypercarnivorous lineages. Not only was *Allosaurus* the iconic form of a particular group of giant sauropod killers—called Allosauroids—but rare specimens have hinted that it got to be just as big as *Tyrannosaurus*. *Allosaurus* was not the milder precursor of prehistory's most famous tyrant, but an agile predator of frightening size and aspect. And,

Allosaurus was the most common big predator in Late Jurassic Utah. Paleontologists are still trying to figure out why this 150-million-year-old carnivore was so abundant on the ancient floodplains it stalked.

(Photograph by the author at the Natural History Museum of Utah)

based on the Cleveland-Lloyd fossils, some paleontologists have speculated that these apatosaur eaters may have hunted in packs.

Since paleontologists started working Cleveland-Lloyd's Jurassic graveyard in 1927, the remains of more than forty-six *Allosaurus* have been extracted from the prehistoric jumble. Most are isolated and scattered in a slurry of osteological remnants; a few crushed bones show that other dinosaurs stepped on the remains of their deceased and defleshed comrades. Paleontologists calculated the minimum *Allosaurus* count based on the number of left femora found at the site, but not every *Allosaurus* buried here contributed the same skeletal element to the bonebed. What's more, only about a third of the bonebed has actually been excavated. There are probably many more *Allosaurus* here than we'll ever know.

Bits of other predators—such as the more massive, knife-toothed carnivore *Torvosaurus* and a particularly big *Ceratosaurus*—have turned up in the same deposit, as well as bones from *Stegosaurus*, the delicately bulky *Barosaurus*, and the blunt-headed sauropod *Camarasaurus*. The early tyrannosaur *Stokesosaurus* and the enigmatic theropod *Marshosaurus* are here, too, and were first discovered in this quarry. But no dinosaur comes anywhere close to the abundance of *Allosaurus*, which is far more numerous than all the other dinosaurs put together.

The yawning *Allosaurus* in the Cleveland-Lloyd visitor center is a composite skeleton that represents the lives of many who died here. And these battered black fossils formed the standard image of what *Allosaurus* was like. Utah's first state paleontologist, James Madsen, Jr., used the Cleveland-Lloyd fossils to catalog every last bone in the *Allosaurus* skeleton in a classic monograph, and *Allosaurus* skeletons based on this quarry's bones can be seen in museums all over the world. When I asked the artist Glendon Mellow to design an *Allosaurus* tattoo for my right arm, he went for inspiration to the Royal Ontario Museum, where an *Allosaurus* skeleton modeled on the Cleveland-Lloyd fossils has stood for years. If you see an *Allosaurus* in a museum, there's a good chance it was based on bones from this isolated spot in Utah.

The buildings that cover the exposed portions of the actual bonebed are a short walk from the visitor center. Dinosaur National Monument it's not, but the BLM has left some fossils in place and installed charcoal-colored casts to re-create an image of just how dense the bonebed is. This is a dinosaurian mess. Skull elements, ribs, vertebrae, and limb pieces are strewn across the exposed rock. The Cleveland-Lloyd Quarry is fossil chaos.

No one knows what created the bonebed, or why *Allosaurus* unquestionably dominates the assemblage. The sheer number and density of carnivores hint that this was a predator trap, not unlike the much more recent saber-toothed cat– and dire wolf–rich La Brea asphalt seeps in the middle of Los Angeles, California. The classic scenario of the Cleveland-Lloyd's history, envisioned by paleontologists like Madsen, goes something like this: During the height of a Jurassic drought, a dehydrated *Stegosaurus* or *Camarasaurus* discovered one of the few ponds on the baked landscape. As the herbivore dipped its head down to drink, its columnar limbs broke through the cracked surface to the sucking mud beneath, and its last hope for life turned into its ultimate doom. There was no way it could escape, and, when it perished, the reek from the victim's rotting flesh attracted opportunistic carnivores for miles around—that is, if its agonized cries hadn't already called in *Allosaurus*. But when those *Allosaurus* tucked into the free meal, they suffered the same fate. Much like an abandoned lobster pot, the putrid pond just kept killing, season after season. Since *Allosaurus* was the most common predator on the landscape, and the overwhelming majority of fossils represent carnivorous dinosaurs, such a scenario would explain the unbalanced collection of bones.

Not everyone agrees that the quarry was a predator trap. The geologic context is ambiguous. Where some see a mucky pond, others see an accumulation of dinosaurs that died in a drought, or a mass of bones washed in from another location. Everyone who studies the quarry has an opinion about what happened here around 150 million years ago. And even within these scenarios, there are other mysteries. If the site was once a predator trap, then why was *Allosaurus* so common while other big predators were so rare? Were *Allosaurus* really that abundant, or did something else skew the sample in their favor?

One possibility is that the *Allosaurus* at Cleveland-Lloyd didn't travel alone. What if the site doesn't represent an accumulation of lone hunters, but a jumble of families or social groups? The Natural History Museum of Utah—the institution that houses much of the Cleveland-Lloyd collection—brought this idea to life when they opened a new paleontology hall in 2011. Stuck in the Jurassic mud, a hapless skeletal *Barosaurus* arches its magnificent neck into the air as an *Allosaurus* family mocks and harasses the giant. One impatient little *Allosaurus* perches on the sauropod's back, digging skeletal claws into flesh that isn't there. The vignette is a brutal Jurassic buffet.

The *Allosaurus* assemblage at Cleveland-Lloyd could be a clue that these dinosaurs worked together to bring down big game. But as far as the evidence goes, there's no way to know for certain. The bonebed was built up over weeks or even years. Dozens of *Allosaurus* are buried at this one place, but why they are all there is a mystery no one has solved. The dinosaurs were buried together, but this doesn't necessarily mean that they lived in rapacious packs. In this case, it's impossible to tell whether the site represents a small number of *Allosaurus* battalions or whether the graveyard is home to many solitary animals. Cleveland-Lloyd is a Mesozoic cold case.

The eastern Utah quarry isn't the only one of its kind. Dinosaur bonebeds found all over the world have raised the possibility that some dinosaurs were social creatures that lived and died together. And if any dinosaur epitomizes the confusing nature of mass graves, it's *Deinonychus*. All you have to do is recall *Jurassic Park* to see why.

While *Tyrannosaurus* embodied brute strength—a force of nature that could break through walls and crush barriers underfoot—*Velociraptor* personified dinosaurian stealth and cunning. However, what we think of as *Velociraptor* was really *Deinonychus*. The name change was thanks to a book published five years before Spielberg's special effects ripped up the celluloid. In his *Predatory Dinosaurs of the World*, published in 1988, the paleoartist Gregory S. Paul imposed his own unique naming scheme on dinosaurs. He decided to lump *Deinonychus*—a sickle-clawed predator John Ostrom named in 1969 from partial skeletons discovered in Montana—with *Velociraptor*, a similar but smaller killer that Henry Fairfield Osborn named in 1924 from bones found in Mongolia. Since *Velociraptor* had been named first, Paul reclassified the larger and different *Deinonychus* as *Velociraptor*. Paleontologists objected to the name change, but, alas, Michael Crichton read Paul's book while he was researching *Jurassic Park*, and he renamed the beefier, more formidable *Deinonychus* in the novel. An actual *Velociraptor* wouldn't have been very threatening. While exceptionally well armed, the predator would have been about the size of a turkey, too small to consider a full-grown human a meal.

What made Crichton's fictionally enhanced raptors so deadly was their intelligence. *Tyrannosaurus* was a single-minded killer, but *Velociraptor* bluffed and feinted to draw victims into their traps. The idea came straight from what the paleontologist John Ostrom hypothesized about the site where *Deinonychus* was originally found. The quarry, tucked away in central Montana, contained the skeleton of a *Tenontosaurus* and several partial *Deinonychus*. The unfortunate herbivore—a member of the ornithischian tribe, cousin to the more familiar *Iguanodon*—didn't have any spikes, plates, or other arrangements. The plain dinosaur had a beaked mouth and a long tail, could have walked either on all fours or on two legs, and seemed like the perfect prey for a predator looking for a relatively defenseless meal. Since Ostrom and his colleagues found several partial skeletons of the carnivore *Deinonychus*—a svelte, muscular predator equipped with retractable toe claws able to sink into the flesh of prey—in the same quarry, it appeared as if a pack of predators had attacked the *Tenontosaurus*. Some perished, but others won out and enjoyed the greasy spoils of their victory. An abundance of *Deinonychus* teeth—shed as the dinosaurs dismembered the *Tenontosaurus*—announced the gruesome victory of the carnivores.

The ability to cooperate immediately set *Deinonychus* apart. Previously, paleontologists had thought that predatory dinosaurs were loners that caught prey on their own and selfishly devoured their spoils. The raptors were different. Spielberg's *Velociraptor* popularized the military tactics of these dinosaurs. Robert Muldoon's last words—before he was mauled and presumably eaten by a *Velociraptor*—were "clever girl," after all.

Yet the quarry Ostrom described showed only that partial dinosaurs were buried together. In 2007, the paleontologists Brian Roach and Daniel Brinkman of the Yale Peabody Museum reviewed the available evidence and determined that the case for pack-hunting *Deinonychus* was not so straightforward. Rather than working together, Roach and Brinkman hypothesized, the dinosaurs competed with each other, and the *Deinonychus* carcasses at the site were individuals killed in the struggle for the meaty *Tenontosaurus*. Naturalists have observed similar behavior among modern Komodo dragons—each lizard is working in its own interest, even as multiple animals gravitate toward the same carcass. In short, competition for fleshy morsels may have killed these *Deinonychus*, not a botched attempt to score a major meal. Much like the *Allosaurus* bonebed and other predatory dinosaur accumulations, the *Deinonychus* graveyard can't be taken as direct evidence that these carnivores acted in coordinated panzer divisions.

While the *Allosaurus* and *Deinonychus* quarries are difficult to interpret, there is at least one bonebed that reflects dinosaur social behavior. This site, tucked away within Canada's Dinosaur Provincial Park, records a horrific catastrophe that killed dozens of dinosaurs at once. The victims were *Centrosaurus*—the iconic horned dinosaur that roamed Cretaceous Alberta around 75 million years ago. In general form, they were typical ceratopsids; *Centrosaurus* had stout bodies with thick limbs for four-footed locomotion. But what distinguished *Centrosaurus* was the ornamentation on the dinosaur's skull. *Centrosaurus* bore a long, slightly curved horn on its nose, lacked brow horns, and had a combination of hooks and hornlets on its frill.

The minimum number of animals—estimated from collected bones—is fifty-seven, but there were undoubtedly many, many more. Based on the sheer number of bones in the quarry, there may have been hundreds of individuals. And while most of the bones that paleontologists have discovered are attributable to adult animals, sub-adult and juvenile *Centrosaurus* were among the casualties, too. The sheer number of fragments and skeletal scraps outlines what must have been one of the most breathtaking sights in prehistory: scores of *Centrosaurus* walking together. There are simply too many animals in one place to deny that the dinosaurs moved as a herd. The site is not a secret *Centrosaurus* graveyard, where animals had slunk off to die over years and years. It's a snapshot of a local disaster.

The bonebed is only where the dinosaurs were buried, and the evidence hints that they died a short distance away. The most popular hypothesis imagines the herd of *Centrosaurus* trying to cross a river, much as caribou, wildebeest, and other herbivorous mammals do today. But something went wrong. Maybe the waters were too high or the animals began to panic, but whatever the immediate cause, much of the herd drowned. The *Centrosaurus* weren't buried right away. Their rotting bodies must have bobbed along and been tossed up onto the riverbank, and there was too much meat for scavengers to dispose of. The local tyrannosaurs, *Gorgosaurus*, ate their fill of the carcasses and left shed teeth behind as evidence, and small tooth marks on some *Centrosaurus* bones put small mammals at the bonanza, too. Eventually, floodwaters rose

and washed the bones of the dinosaurs together, rolling and tum-
bling them along the river bottom until they arrived at their final
resting place. This bonebed didn't come together over many seasons, and can't be construed as a group of dinosaurs squabbling
over the same food source. The site records a quick death for a
massive number of animals that were moving together—a fleeting
vision of how dinosaurs lived and died together.

Bonebeds are one line of evidence that paleontologists draw
on as they examine dinosaur sociality. Some of the best clues
about dinosaur lives come from a different set of clues: footprints.
Admittedly, they're not quite as sexy as dinosaur skeletons. I remember how disappointed I was as a child when I saw tracks left
by dinosaurs that once roamed New Jersey. My parents had es-
corted me to the "Dinosaur Den" of the nearby Morris Museum.
The small, dark hallway featured a big *Stegosaurus*—an armored
giant with no connection to my home state whatsoever—and a
few casts, but the only authentic local fossils were three-toed
tracks left in reddish stone slabs. They just looked like big bird
prints—not altogether surprising, since, before the discovery of
dinosaurs, such footprints in the Connecticut Valley were collo-
quially referred to as "turkey tracks"; the Amherst geologist Edward Hitchcock thought they were traces of moa-like birds. The
footprints didn't speak to me. I wanted real dinosaur skeletons—to see the animals themselves and imagine their fearful
powers.

I didn't understand what dinosaur footprints really were: fossilized behavior. Think of a horned dinosaur—let's say the highly
ornamented *Styracosaurus*—walking across the muddy shore of a
prehistoric lake. As the spiky herbivore trundles along, each foot
leaves an impression in the firm mud, recording—in detail—how
the dinosaur actually moves. Of course, paleontologists don't always know which dinosaur made a specific track. The quirks of
fossil preservation tend to preserve footprints in some places and
skeletons in others, and unless you find a dinosaur literally dead

in its tracks, you can offer only a range of possible candidates. The problem is called the Cinderella Syndrome—matching the right foot with the right fossil "slipper" is really tough to do. Still, dinosaur tracks are distinctive *enough* that paleontologists can narrow down the range of candidates. Scores of sites all over the world have yielded dinosaur highways that reveal patterns demonstrating that they walked, ran, and sometimes traveled together. Some of the tracks found in the Connecticut Valley go in the same direction in parallel—showing that several dinosaurs walked next to each other.

The Connecticut Valley tracks were created by small and medium-sized dinosaurs, most of them theropods. Rare trackways elsewhere show that even huge dinosaurs sometimes ambled along together. The paleontologist Roland T. Bird brought the site to the attention of his colleagues, even if he didn't discover it himself. People who lived in and around Glen Rose, Texas, already knew about the dinosaur tracks when Bird rolled into town in 1938. In fact, rumors of huge dinosaur tracks were what had drawn Bird—who was looking for a set of tracks to lay behind the American Museum of Natural History's *"Brontosaurus"* skeleton—to the vicinity of the Paluxy River in the first place, and there was already a cottage industry based on excavating dinosaur tracks for garden ornaments, of all things. And of course there were forgers, too, some of whom got a kick out of making human-like footprints in the same rock to convince the gullible that people and dinosaurs once walked together.

There was no shortage of footprints around the Paluxy. Some seemed to record the pathways of lone animals, while a special few sites were pockmarked by the movements of multiple dinosaurs. On his maiden voyage to the site, Bird headed in to investigate a tip about some carnivorous-dinosaur tracks on the Davenport family ranch. He found that there were indeed tracks left by a large predator—possibly the ridge-backed relative of *Allosaurus* named *Acrocanthosaurus*—but there was also a field of fossil potholes created by sauropods. Bird considered excavating these, but as he wrote in his journal at the time, "Mrs. Davenport proved to be as big a problem as any in the field." She didn't like the idea of an East Coast scientist swooping in to take tracks off her land. Bird wasn't the first person to ask about them—local track collectors had been pestering the Davenports—and Mrs. Davenport was adamant that the tracks would stay put. "No one had ever removed any tracks from Davenport property to her knowledge," Bird wrote, "and not even the prestigious American Museum of Natural History of New York was about to do so now or at any future date. Any charm exuded by the said museum's field representatives was as so much sweetness and fevered breath wasted on the desert air." Ultimately, Bird appealed to Davenport's sense of curiosity about what sort of animal actually made the tracks to get permission to dig on her land, but that was all: he didn't get the OK to bring any treasures back to New York City.

Bird and his assistants were eager to get to work nonetheless, and they peeled back layer after layer of limestone over an area where a set of sauropod tracks disappeared under the surface. And to his astonishment, there was a wide span of dinosaur tracks that had been hidden under the top layers. "[T]he more we found," Bird wrote, "the more there was left to be found . . . [T]here seemed to be no end to the number nor to the length of the sauropod parade." When the site was fully uncovered, Bird couldn't believe his eyes:

> Here was not a single sauropod trail as I had found on the Paluxy; here a herd of giants had stampeded, or moved, as a single entity. I tried for an accurate count, but it couldn't be done. There were to begin with prints of seven individuals, in a twenty-foot space, but a few feet beyond this well-trodden area the tracks broke down into a hodgepodge. One fact stood out above all; they moved in the same direction, and presumably at the same time.

Not all the footprints were the same size, either. The big and little prints told Bird that teens and adults traveled together. And, since these huge dinosaurs were thought to be swamp-dwelling sluggards at the time, Bird took the general absence of tail impressions to mean that these dinosaurs had been sloshing through shallow water. One set of tracks, in Bird's estimation, had a long drag behind it, and he wondered if this individual was extra large, tired, or "supposed to sign off with his tail." (Bird wrote, "Perhaps scientists of future ages may be able to settle" the question of why dinosaur trackways were not associated with tail drags, as they should have been if anatomical restorations of the time were correct. Bird was merely working within the milieu of his time, and the mystery isn't a mystery at all—dinosaurs did not drag their tails. Bird's observation that "most of my dinosaurs run high-tailed" was a reflection of a reality not recognized at the time he uncovered the tracks.)

The dense accumulation of footprints at the Davenport ranch continued to fascinate paleontologists long after Bird. All those sauropod tracks moving in the same direction were solid evidence that some of these animals traveled in herds. One idea, suggested by the paleontologist Robert Bakker, is that the little ones plodded along in the middle of the group so they would be protected from predators.

But sauropods may not have been such attentive and aware parents. The problem with trackways is that multiple animals, even moving in a single group, often walk over each other's tracks and obscure the trail. In the case of the Davenport herd, composed of at least twenty-three dinosaurs, the little tracks overlap with the big ones, and, according to the track expert Martin Lockley, show that there was no special arrangement to protect the vulnerable young dinosaurs. The smaller dinosaurs simply followed the larger ones, unprotected. The dinosaurs passed through some kind of bottleneck, and so the herd passed through the narrow space in a line rather than as a spread-out group.

Whether sauropods in general formed herds is impossible to say. While these dinosaurs all shared the same body plan, the group was diverse, disparate, and long-lived. It would be silly to assume that such a varied group of animals all behaved in the same way; some were probably solitary, while others plodded *en masse*. Trackways, like the ones in Texas, indicate that some sauropods formed groups at least some of the time, but the precise details of their social structure are almost entirely unknown to us. We know some sauropods were gregarious, but that's almost all the rock record has told us so far.

Trackways are the most direct evidence we have of dinosaur social lives. Even though the idea that *Deinonychus* and other raptors were pack hunters was originally based on flimsy evidence, for example, unique trackways show that deinonychosaurs occasionally traveled together. The peculiar feet of these dinosaurs are the key to their identification. Most theropod dinosaurs supported themselves on three-toed feet, but deinonychosaurs stood on only two, like modern ostriches. Their second toe—the one closest to the midline of the foot—supported their big, retractable claw and was lifted off the ground, creating a very distinctive type of footprint. One trackway made by these dinosaurs records the movement of several individuals in the same direction, spaced about a body width apart—good evidence that this was a social group. And another trackway found in Niger shows where one raptor moved and altered the path of another dinosaur—a subtle moment of early Cretaceous time that gives us another clue that these dinosaurs interacted with each other. Visions of highly organized, socially stratified raptor packs are the products of sensational speculation rather than scientific fact, but at least some forms of these sharp-clawed predators strutted together.

Evidence from trackways and bonebeds undermines the classic image of dinosaurs that I absorbed when I was a child. The dinosaurs I initially encountered were always grouchy-looking, solitary animals—*Stegosaurus* and "*Brontosaurus*" browsed for soft plants on their own, confident that their respective armor plates and giant size would save them, while carnivores like *Allosaurus* and *Tyrannosaurus* were rogues, always lurking behind the next tree. In the books, movies, and museum displays I grew up with, only *Triceratops* and *Deinonychus* were regularly shown in groups, for defense and to better subdue meals, respectively.

Dinosaur Provincial Park's *Centrosaurus* bonebed, the Paluxy tracks, and other sites show that many dinosaurs were gregarious. But not all dinosaur social interactions were friendly. We love to bring dinosaurs back to life to see them tear each other to pieces. From the very beginning, paleontologists have been unable to describe dinosaurs without pondering their impressive powers of attack and defense. In his 1838 treatise on the manifestation of God's power in nature, naturalist William Buckland cast the formidable *Megalosaurus*—which he viewed as an enormous carnivorous lizard—as an intelligently designed killing machine. In fact, Buckland proclaimed, *Megalosaurus* was so well suited to killing that the efficiency of the predator actually "tend[ed] materially to diminish the aggregate amount of animal suffering. The provision of teeth and jaws, adapted to effect the work of death most speedily, is highly subsidiary to the accomplishment of this desirable end."

Even herbivores were restored as vicious brutes. In one of the earliest dinosaur paintings—composed by the celebrated painter of biblical catastrophes John Martin—a serpent-like *Iguanodon* opens a mouth full of jagged teeth to bite an attacking *Megalosaurus* in protest. The art looks less like a scientific restoration than like a medieval vision of wyrms or wyverns that have ensnared each other, millions of years before there was a St. George to slay them.

While not quite as apocalyptic as what Martin envisioned, a single bone at the Utah State University Eastern Prehistoric Museum records a traumatic battle between *Allosaurus* and *Stegosau-*

rus. The small museum squats along the main drag in Price, not far from the Cleveland-Lloyd quarry. A high-kicking, featherless *Utahraptor* in bronze greets visitors entering the parking lot, and a skeletal reconstruction of the same dinosaur mimics the ninjutsu pose inside the museum's foyer. The last time I visited, after my failed first attempt to visit the nearby dinosaur boneyard, many of the other skeletons were off display, scheduled to have their tails lifted and spines readjusted to fit the twenty-first-century dinosaur bauplan. But there was still plenty to see in the gallery, including an *Allosaurus* vertebra with a *Stegosaurus* spike driven through it.

The fossil wasn't actually found like this. The original *Allosaurus* tailbone—uncovered at the Cleveland-Lloyd bonebed—had a weird, C-shaped puncture along one side. According to the museum director Kenneth Carpenter and his colleagues, a *Stegosaurus* spike fits snugly in the wound. (Inspired by one of Gary Larson's *The Far Side* cartoons, Carpenter has also proposed that a stegosaur's array of tail spikes should be called a "thagomizer.") The pathology is a vestige of a Jurassic fight, and the *Stegosaurus* even left part of itself behind. The *Allosaurus* bone didn't heal properly—a clue that the tip of the *Stegosaurus* spike broke off and was left embedded inside the carnivore's body. This might have been a frequent hazard for both the attackers of *Stegosaurus* and the spiky dinosaur itself. About 10 percent of *Stegosaurus* tail spikes show healed breaks at their tips. Most often, the researchers suspected, *Stegosaurus* spikes would have slashed open long wounds on the sides of would-be predators, but, if the carnivore approached from a particular angle, the spikes would have been more likely to break off and lodge inside the assailant's body. Even though the actual event is lost to prehistory, the fact that *Allosaurus* and *Stegosaurus* faced off is preserved in bone.

Members of the same dinosaur species fought each other, too. Healed skull wounds, like those on the young *Tyrannosaurus* "Jane" at the Burpee Museum of Natural History in Illinois, show that tyrannosaurs fought by biting each other right on the face. Nor was intraspecies bickering the sole province of carnivores. *Tricer-atops* bore signs of their battles on their skulls.

While passing through Claremont, California, in the fall of 2011, I stopped by the Raymond M. Alf Museum of Paleontology to visit my friend, the paleontologist Andy Farke. The main reason for my visit was to see the museum's new paleontology hall, and I also wanted to check out one of Farke's *Triceratops* models. A few years before, Farke had used a pair of dinosaur sculptures to visualize how the famous ceratopsians locked horns. As he led me to his office, I asked Farke if I could see the polyresin sources of his inspiration. He gladly handed the model to me from its spot on the shelf: a detailed sculpture of a *Triceratops* skull roughly 15 percent of actual size. Matching it with an identical model, Farke staged a mock battle to show how *Triceratops* might have fought.

As explained in countless dinosaur tomes, *Triceratops* had two long horns over the eyes, a short horn on the nose, and a broad, solid frill (or, if the "Toroceratops" hypothesis is upheld, at least solid for most of its life). To generations of paleontologists, as well as to kids in sandboxes in New Jersey, these adornments looked like weapons honed for combat. When the Yale University paleontologist Richard Swann Lull described a pathological specimen of this dinosaur—which might be called *Nedoceratops* if it turns out to be a distinct genus—he wrote, "the supraorbital horns are the sole aggressive weapons while the widely expanded frill served admirably to withstand the shock of the adversary's horns. We have here a precise analogy with the knight of old tilting with his spear and shield." Whether goring a *Tyrannosaurus* or fending off a rival, *Triceratops* surely used the weapons to jab, block, and parry.

No one actually investigated the defensive abilities of *Triceratops* in detail. Its defensive prowess seemed self-evident. But Farke wondered how the dinosaurs would have jousted. When he manipulated the two models to see how *Triceratops* fought, Farke found only a few possible horn-lock positions. *Triceratops* could have angled their heads so that only one brow horn of each individual hooked around the other's; they could have tilted their heads even further down so that both brow horns locked around each other's; or they could have offset their heads to the side so that their brow horns locked and their nasal horn jutted up into the frill.

Farke wasn't just fooling around with toys. By determining the range of horn-locking positions, he could look to actual *Triceratops* skulls for signs of combat. If the dinosaurs were fighting in the ways Farke predicted, then the parts of their skulls the horns scraped against should have been damaged. Farke followed up on his proposal with colleagues Ewan Wolff and Darren Tanke in a 2009 paper, "Evidence of Combat in *Triceratops*." The paleontologists looked at the different parts of the skull to see if there was any clear sign of regular combat. As anticipated by Farke's models, the lower bones on the outside of the *Triceratops* skull—the squamosal and the jugal, or cheekbone—had the highest incidence of injury.

The damage on the *Triceratops* skulls was consistent with the idea that the animals often locked horns, and the pathology counts were significantly higher than the occurrence of lesions in the distantly related *Centrosaurus*. Within the sample of this dinosaur—an opposite of *Triceratops*, with a long nasal horn and short brow horns—a few individuals had lesions on their cheekbones, but the pattern of injury wasn't the same. These dinosaurs were clearly doing something different. Indeed, *Centrosaurus* did not have a set of horns fit for interlocking, so, Farke and coauthors speculated, maybe these dinosaurs relied on visual displays while competing. This might in fact have been the predominant rule among ceratopsids. Ceratopsians had a fantastic variety of horn arrangements and frill shapes, many of which looked ill-suited for combat—either with rivals or with predators. In the recently described *Kosmoceratops*, for one, the short frill features a row of short horns folded over forward, and the dinosaur's facial ornaments consist of two short horns jutting out sideways and a low, almost bladelike nasal horn. If these dinosaurs stabbed and battered each other with their heads, I have no idea how they did it.

Indeed, many dinosaur ornaments probably had more to do with display than with defense. Dinosaur fashion was clearly focused on the bizarre: horns, spikes, plates, sails, and the like were common features, though each species wore them in a slightly different way. And for a long time, the purpose of much of this ornamentation seemed clear: the horns of *Triceratops*, the armor of *Ankylosaurus*, the domed head of *Pachycephalosaurus*, and the plates of *Stegosaurus* were defensive structures used to wound predators and, presumably, to win contests with rivals of the same species. But the way we think dinosaurs behaved is constrained by our imagination, or lack thereof. Just because an anatomical structure looks like a lance, a mace, or a battering ram doesn't mean it was used as such. When scientists began to investigate the properties of the dinosaur arsenal, the difference between defensive weapon and display structure was obscured.

Of all the dinosaurs with odd body structures, the thick-skulled pachycephalosaurs are among the most perplexing. These weren't quadrupedal bruisers, but bipedal herbivores with domes of bone and accessory spikes on the tops of their heads. They seemed perfectly adapted for running headlong at each other and cracking skulls, much as bighorn sheep do today to determine dominance in their social groups.

The most famous of all dome-heads, and the dinosaur from which the group derives its name, is *Pachycephalosaurus* itself. In profile, the dinosaur's head was wedge-shaped, with a narrow nose decorated with bumps, an expanded and rounded skull ruff, and a series of bumpy ornaments around the back edge. Why any dinosaur should have such a head was a mystery, but, following the logic that bizarre structures were often for attack or defense, Edwin Colbert proposed that these dinosaurs frequently butted heads in competition. What else was a thick skull roof for?

The pachycephalosaur-ram analogy was a rough one. The

modern mammals have curved horns that contribute to a wide surface suited to absorbing shock. *Pachycephalosaurus* had a reinforced and rounded dome. When I visited the pachycephalosaur expert Mark Goodwin at Berkeley, I mused that contact between two such skulls would be like smacking two bowling balls into each other: all the force would be concentrated on one small area of contact where the domed surfaces met. "That's right!" He laughed. If the dinosaurs ran at each other, they could fatally injure themselves.

I wasn't the first person to notice the drawback. Several paleontologists have suggested that pachycephalosaurs would have been rather poor head-butters, and Goodwin's work with Jack Horner has underscored how lousy the structures were for going head-to-head. As the paleontologists found with other dinosaurs, the skulls changed dramatically as they aged, and the bone structure thought to make for good shock-absorbing powers was just a transitory state that disappeared before the animals reached full maturity. In Goodwin and Horner's estimation, pachycephalosaur domes were probably signals to help members of a species recognize each other, and may have played a role in sexual competition and selection come mating season. The dinosaur "weapons" turned out to be social signals. Indeed, the weird ornaments among dinosaurs undoubtedly featured in their social lives, whether used for identifying members of their own species, for attracting mates, or for intimidating rivals. Of course, there were some cases—like *Triceratops*—where defense and display coincided, and a recently discovered fossil hints that this might have been true for *Pachycephalosaurus*, too.

For years, critics of the head-butting pachycephalosaur idea noted that no one had ever found the kind of damage on these skulls that you'd expect if these dinosaurs were ramming each other. If the dinosaurs used their heads as weapons at all, they probably butted each other in the flanks. But in 2012, Joseph Peterson and Christopher Vittore identified a *Pachycephalosaurus* skull that had been injured by a traumatic impact and had suffered an infection following the incident. Peterson and Vittore concluded that head-butting behavior was the most probable explanation for the damage. As paleontologists look over their collections of pachycephalosaur skull domes, perhaps other examples will show up and fuel the ongoing oramentation debate.

Pathologies like damaged frills and dented skull domes record fleeting interactions between dinosaurs, but the fact that the structures evolved at all can tell us something about dinosaur social lives. Horns, frills, domes, spikes, plates, and other features were multifunctional: they could be used for battle, and acted as visual signals, too. They were symbols that dinosaurs used to communicate, and, like bonebeds and tracksites, they offer us clues about how dinosaurs interacted with each other. Indeed, co-option is what evolution thrives on, and what a structure is currently used for doesn't always reveal how that trait first evolved. Feathers are a perfect example. Even though many dinosaur ornaments were bony, feathers and their rudimentary, fuzzy precursors were a widespread dinosaur feature that undoubtedly played some role in dinosaur society. Indeed, even though plumage eventually allowed some dinosaurs to take to the skies, feathers originally evolved for other reasons related to display and insulation, and were only later repurposed for flight. Thanks to a fortuitous discovery, we can now start to understand how dinosaurs may have used color—even bright, gaudy color—to show off.

Seven

Dinosaur Feathers

Sometimes I get a little selfish about dinosaur skeletons. As thrilled as I am that museum dinosaur exhibits are so well attended, the stampeding hordes of schoolchildren and waves of parents pushing their stroller-bound kids through narrow exhibit pathways can be more than a little agitating. Walking through dinosaur displays at peak hours requires serious agility to avoid the swarms of little ones buzzing around the place. And that's not to mention the fact that few people seem to read the museum labels—any sharp-toothed predator is a *Tyrannosaurus,* and every supersized sauropod is a *"Brontosaurus."* I want to butt in and point out the correct names, but when I've done so, I have often been met with annoyed glares. Better to keep my mouth shut and let the families enjoy their time in the midst of the fossilized superstars. "Be nice," I have to remind myself, ". . . you're just one of those irrepressible dinosaur fanatics all grown up."

I often watch the tide of visitors go by from the bench at the Natural History Museum of Utah's paleontology lab. Behind a set of high glass windows, the other volunteers, technicians, and I go to work in a scientific fishbowl among tables stacked with fossils and covered in flecks of prehistoric rock. Sometimes I'll be absorbed in my work—breaking off tiny pieces of sandstone from a fossil in the raw—and over the whine of the air-powered scribe I use to pick away at the encasing rock, I'll hear a bang on the windowpane as a gaggle of kids catapults themselves onto the glass to get a better look. They're so excited—until they realize that cleaning dead dinosaurs is a real pain in the ass, a war of millimeters between you and the matrix that surrounds the fossil bone.

On some afternoons, when the flow of museum patrons has ebbed, I take a few minutes to amble through the exhibit halls. The quiet of the vast, dim space reminds me of my first trip to see New York City's grand dinosaurs. The osteological galleries are among the few places where I can tune out the various distractions, always just a tap away on my smartphone, and let my mind drift as I walk past a pack of *Allosaurus* poised on tiptoe and gaze up to the ludicrously long neck of the museum's titanic *Barosaurus*. I feel at home among the dinosaurs.

And in those moments, I can't help but wonder what the animals looked like when they were alive. Dinosaur skeletons are beautiful, exotic frameworks that supported flesh in life, and are the jumping-off point for my daydreams now. Fossil impressions of pebbly dinosaur skin fill in some of the details, but that's just the canvas. Dinosaur color is another matter altogether. I can imagine sloshing buckets of polka-dot paint over the museum's many-horned *Utahceratops*, but I doubt that in reality he would have looked so conspicuous. On the other hand, the traditional garb of drab green or gray isn't very appealing, either. Maybe the horned dinosaur shared a palette with today's African antelope, like the bongo—sienna shades set off with patches of black and thin white stripes. I can always revise the color scheme later.

When I was a kid, books and museum displays told me that dinosaur color was one tantalizing aspect of *Apatosaurus* and company that we'd never be able to find out. The mystery was as frustrating as it was fascinating, and, from what I've heard, "What color were dinosaurs?" is still the question paleontologists field most often. For a long time, there was no answer. Whether working in paint or with the animatronic dinosaurs that terrified me the first time I saw them, artists could have free rein to pick any color scheme they wanted without fear of scientific reprisal.

I used this to my advantage when I was still a young dinosaur fan and created a few dinosaur drawings for the paleontologist Peter Dodson. My father told me he was taking me to Dodson's lecture at the local library, and I couldn't wait. This was my chance to impress a real paleontologist! Someone who could open doors to fantastic collections and fossil-rich field sites! So I spent the afternoon sketching dinosaurs, including what turned out to be an atrocious drawing of the many-horned dinosaur *Styracosaurus*. This dinosaur had the same build as *Triceratops*, but with a vastly different head—a long nasal horn, short brow horns, and an array of intimidating spikes jutting backwards from its frill. And I honored this proud dinosaur by giving it a truly awful color scheme, too. The ceratopsid's beak reminded me of a macaw, so I colored the dinosaur fire-engine red with a splash of white and black around the eye. I started with the eye first, and instantly regretted it. All the same, who could say? Later that night, I presented Dodson with the garish dinosaur. I'm forever grateful that he didn't burst out laughing.

That dinosaurs might have been so fantastically colored was a relatively new idea during my childhood in the 1980s, a concept that grew out of the notion that dinosaurs were more birdlike than anyone ever expected. Before that, dinosaurs traditionally wore stately, subdued colors. Olive green and mud brown were the default choices. Even movie dinosaurs, who were meant to be ferocious, vibrant creatures, had scaly hides duller than a pet-store lizard. The comically carnivorous *"Brontosaurus"* in *King Kong* (as well as the rest of Skull Island's Mesozoic fauna, for that matter) flickered as gray monstrosities in weekend reruns of the film on my family's television set, the grayscale colors a necessity of the early days of cinema. But dinosaurs in the age of color were lackluster, too. Ray Harryhausen's anachronistic *Triceratops* and *Ceratosaurus* in 1966's *One Million Years B.C.* wore uniform shades of brown and gray, and the brontosaur family of *Baby: Secret of the Lost Legend* were solid charcoal. Even *Jurassic Park* (which debuted two decades after artists and scientists took the colorful lessons of the Dinosaur Renaissance to heart) featured typically drab dinosaur stars. Apparently Steven Spielberg wanted classic Hollywood monsters rather than the most accurate dinosaurs science could offer. Jack Horner, who has been a paleontology consultant for blockbuster dinosaur films, once told me that the director drew a hard line on what the dinosaurs should look like, noting that Spielberg felt he couldn't "scare people with Technicolor dinosaurs."

By the time *Jurassic Park* came out, the dull dinosaurs were behind the times. The realization that dinosaurs were extremely active, birdlike creatures opened a world of color possibilities to dinosaur artists. And some of those paleo-illustrators have had no trouble going overboard: think *Deinonychus* draped in neon colors, like a Cretaceous Cyndi Lauper. For the most part, though, artists turned to the natural world around them for some clues about dinosaur color. The paleoartist Gregory S. Paul, in his classic book *Predatory Dinosaurs of the World*, laid out a few rules for shading dinosaurs. "Since big living reptiles, birds, and mammals are never gaily colored like many small reptiles and birds," Paul wrote, "one can assume that subdued colors were true of the big predatory dinosaurs, also, which to human sensibilities gives them a dignified air appropriate to their dimensions and power." Stripes, spots, or patches of iridescent color around the snout are acceptable, Paul said, but duller color schemes are the most practical.

But dinosaur color is no longer strictly the realm of speculation and artistic taste. Living dinosaurs, as well as fossils bearing impressive plumage, have provided an unprecedented window into prehistory. The key to the whole puzzle is a simple, beautiful fact that has irrevocably changed the way we look at dinosaur lives. It is simply this: birds are dinosaurs. It's a strange notion to think that the little hummingbirds that come to sip from the feeder planted just outside my window are part of the sole surviving dinosaur lineage, but there's no doubt about it: the Age of Dinosaurs continues. Birds just so happened to be the one dinosaur lineage that survived the end-Cretaceous extinction. It took more than a century for scientists to agree on this point, and it's worth taking a moment to consider the long history of the debate and how it relates to what our extinct dinosaur friends looked like.

There has always been one critical fossil that comes up in the discussions paleontologists have about the origins of birds: *Archaeopteryx*. Described in 1861 from a feather and a partial, feathery skeleton discovered in a German limestone quarry, this mosaic of reptilian and avian traits has been the keystone for varying theories about how birds originated. Lately, a slew of dinosaurs with plumage has led paleontologists to question what *Archaeopteryx* really was.

I remember exactly where I was when *Archaeopteryx* was threatened with demotion from its place as an evolutionary icon. I was sitting at an Exxon station in the middle of nowhere Montana, waiting for my rented SUV to finish fueling so I could continue my journey from the isolated town of Ekalaka (where I had been looking for dinosaurs with the paleontologists Thomas Carr and Scott Williams and their field crews) down to Thermopolis, Wyoming. After running into the convenience store to buy the requisite snacks and caffeine for my seven-hour trip, I checked my messages to see if I had missed anything important while I was in the field. New dinosaur studies come out faster than you might imagine.

E-mails trickled into my inbox. Mostly junk. But then there was a spate of messages from the ever-prolific Dinosaur Mailing List, titled "Greg Paul is right (again); or 'Archie's not a birdy.'"

The title referred to an idea, suggested years ago by paleoartist Paul and others, that *Archaeopteryx* was not the earliest known bird, but in fact one of a *variety* of feather-covered dinosaurs more closely related to the famous predators *Deinonychus* and *Velociraptor*. The idea had been kicked around over the years without much enthusiasm, but a paper in *Nature* had been released that afternoon which shook up the bird family tree and punted *Archaeopteryx* off to the non-avian dinosaur branch.

I cursed my luck that I couldn't get the report at my roadside stop, but since I was the only one at the pumps, I didn't feel bad about taking a few extra minutes to see what news services were saying about the theory. If there's anything reporters love more than a story about *Tyrannosaurus rex*, it's a story claiming that some facet of dinosauriana we had taken for granted has turned out to be wrong.

The splash of articles on the study didn't disappoint. "'Oldest bird' Archaeopteryx knocked off its perch in controversial new study," said one. Another baited evolution denialists with the title "Newly discovered dinosaur could disprove 'earliest bird' theory," although the article itself only stumbled through a litany of tidbits about Archie and a new feathered dinosaur dubbed *Xiaotingia*.

Apparently, after analyzing the evolutionary relationships of *Xiaotingia*, the paleontologist Xu Xing and colleagues found that both *Xiaotingia* and *Archaeopteryx* were more closely related to feathered but non-avian dinosaurs like *Velociraptor*. Bizarre, poorly understood forms such as *Epidexipteryx*—a small theropod decorated with ribbon-like feathers, with a mouth full of procumbent teeth—fell out closer to the ancestry of birds in this new evolutionary tree.

Depending on how you look at it, this was either a case of the best or worst possible timing. The entire reason I was on the road to Thermopolis—a tiny dot in the middle of Wyoming, best known for its hot springs—was to see the only *Archaeopteryx* specimen in the United States. If the report held true, the *urvogel* (original bird) had been cast down just a few hours before I was due to roll into town. "You've got to be kidding me," I thought as I pulled out of Exxon and started my long interstate drive.

Now, every *Archaeopteryx* specimen ever found—from a single isolated feather used to establish the creature's name in 1861 to the eleventh specimen announced in 2011—has come from southern Germany. The one I was going to see was one of the more recent discoveries, but we'll get to that in a moment. All the *Archaeopteryx* skeletons are preserved in limestone slabs that record the Jurassic life that sank to the bottom of an ancient sea that covered much of Europe around 150 million years ago. Crustaceans, fish, pterosaurs, small dinosaurs, and other creatures have all turned up in quarries, but the most cherished of all the fossils are those of *Archaeopteryx lithographica*. The high-definition preservation of these fossils not only recorded the anatomy of the creature's bones, but, in many of the specimens, vestiges of the feathers, too. That's what made the first *Archaeopteryx* skeleton ever found such a sensation.

Known as the "London specimen," the animal resembled certain dinosaurs in terms of its anatomy, yet *Archaeopteryx* clearly had feathers. Freshly embroiled in the controversy stirred by Charles Darwin's *On the Origin of Species* in 1859, Victorian evolutionists privately rejoiced that the creature was a confirmation that transformations from one kind of creature to another were actually possible. As the paleontologist Hugh Falconer called it, in a private letter, *Archaeopteryx* was a "strange being *à la* Darwin," and Richard Owen (who obtained the first skeletal specimen for what is now London's Natural History Museum) deemed *Archaeopteryx* to be the "by-fossil-remains-oldest-known feathered Vertebrate" and the earliest known bird.

Owen's ambitious plans for his museum were what brought *Archaeopteryx* to England. He wanted unique, dazzling fossils for his collection, and convinced the museum to front the cash for the

German fossil. Once everyone understood how important the early bird was German paleontologists were sore that their country's prize fossil had been so easily acquired by foreign scientists. While the second *Archaeopteryx* skeleton—called the "Berlin specimen," the most beautiful fossil of all time—was almost sold overseas to O. C. Marsh at Yale, and the cryptic Haarlem specimen—confused for a pterosaur until 1970—is held at the Teyler Museum in the Netherlands, all but two *Archaeopteryx* stayed in Germany. If you see an *Archaeopteryx* in an American museum, chances are that you're looking at a *cast* . . . unless you're in the middle of Wyoming.

Going by appearances alone, you'd never guess that Thermopolis contained anything as important as an *Archaeopteryx*. Faded signs along the highway leading to the isolated town give equal billing to the Wyoming Dinosaur Center and the "Safari Room"—a dining room decorated by the stuffed spoils of a big game hunter at the town's overpriced Days Inn. You know you're getting close to the local dinosaur showroom when you spot a metal *Allosaurus* skeleton on a street corner along the main drag, frozen as if roaring at the cars passing by.

I follow the suburban streets to the gravel parking lot outside the museum, anxious to get out of the sun and into the cool building where the famous fossil rests. The exterior of the Wyoming Dinosaur Center is as mundane as the drab dinosaurs I met in elementary school. There are no windows, columns, statues, or, really, much of anything. The gray building displays "Wyoming Dinosaur Center" in mismatched shades of green, and the whole structure baked in the heat of the August afternoon. I pay my ten-dollar entry fee and am directed by a disaffected young woman to a corridor that will lead me through the exhibits.

Contrary to its title, the Wyoming Dinosaur Center displays a variety of other forms of prehistoric life. The dinosaurs are the real draws, of course, and keep people moving along the hallway, past the petrified invertebrates and fossil fish. Along the way, I notice one large slab to the left of the path, depicting an aggregation of pancake-size ancient horseshoe-crab-like arthropods called trilobites; a nearby shelf displays a reproduction of the wormlike, schnozzle-faced invertebrate called a Tully monster (once a contender for the identity of the Loch Ness Monster, in fact); and a small alcove presents an array of early tetrapods, the amphibious vertebrates that were the first to clamber onto land around 375 million years ago. And then there are the dinosaurs. Some of the fossils on display are authentic. Others are casts, which isn't too surprising, given how difficult it is to put together heavy, invaluable bones of prehistoric creatures.

I didn't come for fiberglass dinosaurs. What I had driven all morning to see was the real thing, and there it was. Set behind a protective pane of glass, the Thermopolis *Archaeopteryx* rests in its limestone tomb. The skeleton, about the size of a raven's, was preserved in an odd pose, presenting the dinosaur as though it had fallen backwards off a bicycle—legs splayed, head thrown back, arms to the side, and all surrounded by the faint impressions of feathers. The little dinosaur's skeleton resembles the fierce anatomy of *Velociraptor*, but the array of feathers gives the *Archaeopteryx* fossil a subtly different character. I stand and stare at the fossil for a while, tracing its form along the slender toes and thin legs up the contorted spinal column to the animal's wishbone, still situated between the birdlike shoulders. A heavyset man and his towheaded son, both decked out in the logos of their favorite sports teams, slowly amble past and don't pay the little slab much attention. The dramatic scene of a skeletal *Monolophosaurus* sinking its recurved teeth into the side of a long-necked *Bellusaurus* is apparently far more interesting and consistent with the character of the "terrible lizards."

They have no idea what they are missing! As I daydream about the bones, I wonder how this fossil wound up in such an isolated little town. Outside of Germany, I would have expected such a fossil to be on display in one of the venerated institutions further east—Chicago's Field Museum, the American Museum of Natural History in New York City, or Pittsburgh's Carnegie Museum of Natural History. What the hell was *Archaeopteryx* doing here?

It turns out that no one knows when this specimen was originally collected or where it was found. Rumor has it that the fossil was discovered some time in the 1970s, and the specimen was effectively a private secret until 2001, when a Swiss collector's widow offered it for purchase to Germany's Senckenberg Museum in Frankfurt. The museum declined, but in 2005 Burkhard Pohl of the Wyoming Dinosaur Center arranged a deal whereby the *Archaeopteryx* would be on long-term loan to the private museum. And even though fossils receive some protection in most federal states of Germany under Monument Protection Acts, Bavaria (where the *Archaeopteryx* fossils are found) doesn't have such a law, and so the export of the *Archaeopteryx* to Switzerland, and later to the United States, was perfectly legal, no matter how painful it was to see the specimen wind up at a commercial institution far from home. Too many countries have been robbed of their prehistoric heritage thanks to lax fossil regulations.

Had I visited the museum a day earlier, I wouldn't have given a second thought to what I was looking at. I would have taken it as current fact that, as it had been regarded for a century and a half, *Archaeopteryx* was the key to bird origins. Whether or not *Archaeopteryx* was a direct ancestor of later birds didn't matter—as the earliest bird, the feathered dinosaur represented the form of the very first avians. But now I had to wonder about the nature of the creature. Was the *Archaeopteryx* behind the glass truly an early bird, or a different kind of dinosaur simply hiding behind beautiful plumage?

I also knew that *Archaeopteryx* has always had a controversial place in our developing understanding of how birds evolved. Even around the time the fossil was originally discovered, and Richard Owen asserted that the bird lineage started with such a creature, other naturalists were not so sure. Darwin's friend and vociferous defender Thomas Henry Huxley sidelined *Archaeopteryx* as a weird animal that was almost entirely irrelevant to the question of bird origins. Instead, an influence on the evolutionary circumlocutions of German biologist Ernst Haeckel, Huxley proposed that the origin of modern birds went through a three-step process, starting with creatures similar to the small dinosaur *Compsognathus*, a diminutive theropod found in the same deposits as *Archaeopteryx*. "There is no evidence that *Compsognathus* possessed feathers; but, if it did, it would be hard indeed to say whether it should be called a reptilian bird or an avian reptile," Huxley wrote.

Contrary to what has so often been claimed on his behalf, Huxley didn't suggest that birds evolved directly from any known dinosaur, but proposed that something in the general form of *Compsognathus* was adapted into a flightless bird akin to an ostrich or an emu, and that these birds were the ancestors of flying birds. *Archaeopteryx* was just an evolutionary sideshow that illustrated that birds could possess reptilian traits, but did not fit anywhere into Huxley's scheme.

True to the often contentious nature of science, not everyone agreed with Huxley's proposal. Paleontologists such as Samuel Williston, Franz Nopsca, and O. C. Marsh hypothesized that birds really did have a direct dinosaurian origin. Exactly which dinosaurs was the real matter of debate. Some authorities favored the small, generally birdlike theropod dinosaurs, while others suggested that ornithischian dinosaurs such as *Hypsilophodon*— on the basis of their birdlike hips—were the true ancestors of birds. Still other naturalists mixed and matched these ideas. Perhaps some birds evolved from one dinosaur group, while the rest were derived from the other. Then again, Richard Owen and Harry Govier Seeley insisted that birds had evolved from pterosaurs, a different kind of archosaur that flew thanks to membranes stretched over an elongated finger. Huxley and other naturalists disputed this—the characteristics that united birds and pterosaurs were instances of convergence related to a similar lifestyle—but no one knew for certain exactly how birds evolved. And, despite Huxley's difference of opinion, *Archaeopteryx* became the singular touchstone for understanding the transition from reptile to bird. Any theory of bird origins had to take *Archaeopteryx* into account.

Even as paleontologists agreed that *Archaeopteryx* was the earliest bird, though, they were left with the question of what sort of reptile it had evolved from. The Scottish paleontologist Robert Broom suggested a solution in 1913 that made sense of the traits shared by dinosaurs, pterosaurs, *Archaeopteryx*, and other birds. Before the era of the pterosaurs and dinosaurs, during the earliest parts of the Triassic, the crocodile-like archosaurs ruled. One of these creatures, *Euparkeria*, was a bipedal, carnivorous croc relative that was old enough and generalized enough that it could be a common ancestor for dinosaurs, pterosaurs, and birds. If all three lineages evolved from such a creature—a common and relatively unspecialized rootstock—then that would explain why they were so perplexingly similar to each other.

It wasn't until an early-twentieth-century artist took up the question of bird origins that the answer was considered settled. Gerhard Heilmann was an accomplished illustrator as well as an amateur paleontologist, and in 1926 he published an English translation of a series of articles he had written in Danish called *The Origin of Birds*. I was fortunate enough to track down a copy a few years ago, and the book is a real treasure. The glossy pages are filled with detailed comparative drawings of bird and dinosaur skeletons, and Heilmann illustrated a few dinosaurs in active poses, such as a pair of *Iguanodon* sprinting over the Cretaceous plains. Heilmann's scientific argument was just as elegant as his drawings. Even though he acknowledged that some dinosaurs were birdlike, there was one feature that in his view barred dinosaurs from bird ancestry. Or rather, it was the lack of a feature. Heilmann knew that birds have a wishbone, or the modified set of

clavicles known as a furcula. As far as Heilmann knew, no dinosaur had ever been found with these bones. Dinosaurs had apparently lost their clavicles during the course of evolution, and since
a feature couldn't re-evolve once it had been lost, Heilmann reasoned, there was no way that dinosaurs could be ancestors of birds.
The next closest group that had clavicles contained *Euparkeria* and
its croc-like kin, and so Heilmann concluded that birds and dinosaurs had so many features in common because they had evolved
from a common ancestor.

Paleontologists found Heilmann's argument very persuasive—
so much so that they overlooked the fact that dinosaurs did indeed
have clavicles! A wishbone can clearly be seen in a diagram of
bones published with the description of the beaked theropod *Oviraptor* in 1924, and a wishbone was found among the bones of the
small theropod dinosaur *Segisaurus*, described in 1936 from a skeleton found crouched in a birdlike, roosting position. Heilmann's
hypothesis had become so entrenched that paleontologists somehow missed even seeing these clavicles, and the idea that birds and
dinosaurs independently evolved from a common, crocodile-like
ancestor remained in favor—until a sharp-clawed dinosaur cut
through the debate.

In 1969, the Yale paleontologist John Ostrom named *Deinonychus antirrhopus* from a quarry full of partial skeletons in Montana.
With grasping hands, a long, still tail, and, most remarkable of all,
a hyperextendable toe capable of plunging the dinosaur's "terrible
claw" into prey, this dinosaur was clearly an agile and active predator. *Deinonychus* seemed as different as could be from the traditional
vision of idiotic, swamp-bound dinosaurs—like the ones Ostrom
himself had helped design for the Sinclair pavilion of the 1964
World's Fair—but the osteology of this dinosaur was not totally
unprecedented. *Deinonychus* was very birdlike, and Ostrom quickly
recognized the similarity between his newfound predator and
Archaeopteryx. The dinosaurian origin of birds had clawed its way
back into the scientific spotlight.

•

The idea that birds are dinosaur descendants changed our entire perception of what dinosaurs were. If modern birds are dinosaurs, and dinosaurs resembled avians, then long-held assumptions about dinosaur biology had to be wrong. Maybe not all dinosaurs hopped around like magpies or ran with the grace of an ostrich, but the links between *Archaeopteryx* and *Deinonychus* hinted that some bird traits—such as highly active metabolisms, warm body temperatures, and even feathers—originated deep within the dinosaur family tree.

A 1975 article by Bob Bakker, one of Ostrom's students and the guy who catalyzed the Dinosaur Renaissance, included a restoration of the Triassic dinosaur *"Syntarsus"* with feather-like scales and a crest of plumage on its head as a speculative tribute to the revamped avian dinosaur hypothesis. And, Bakker noted, such a view generated "a particularly happy implication" for dinosaur fans: "the dinosaurs are not extinct; the colorful and successful diversity of the living birds is a continuing expression of basic dinosaur biology."

Ostrom's and Bakker's ideas filtered through to the documentaries I eagerly watched in my youth. One of my favorite shows was *The Dinosaurs!* on PBS. (Documentaries about the prehistoric celebrities in the late 1980s and early '90s regularly combined the word "dinosaur" with whatever number of exclamation points was desired to make their point, from *Dinosaur!* to *The Dinosaurs!* and the extra-emphatic *Dinosaurs! Dinosaurs! Dinosaurs!*) One Thanksgiving Day, PBS ran the entire four-part series in a dinosaur marathon, giving me hours of prehistory-fueled joy while the traditional holiday dinosaur, dressed and stuffed, was downstairs in the oven. In one episode, which highlighted the essential connection between dinosaurs and birds, a little green dinosaur—*Compsogntathus*, I presumed—ran through an ancient forest. As the chicken-legged beast climbed up a log, though, it quickly sprouted feathers and took on more of a confident strut, all before leaping into the air and metamorphosing into a modern pelican.

An episode of PBS's series *The Infinite Voyage* included a little more detail. A very fluffy *Deinonychus* went transparent, showing key bones in the skull, arms, hips, and legs, and as the dinosaur ran it transformed into an *Archaeopteryx* and, ultimately, took flight as a crane. On the outside, a modern bird and something like *Deinonychus* might seem drastically different, but when you look at their skeletal framework, the differences aren't so extreme, after all.

Despite all this conditioning, I still thought feathered dinosaurs looked silly. Dinosaurs were supposed to look mean and scabrous. With feathers on, *Velociraptor* just looked like a big chicken. Plush, downy dinosaurs in gift shops did nothing for me. They looked far too cuddly to be adept flesh-renders. *Jurassic Park* entrenched visions of olive-green, scaly carnivores in my young mind, and even now, there are some absolutely daffy feathered dinosaurs that I feel downright embarrassed for. One of the worst models is on display in Las Vegas—a *Deinonychus* plastered with feathers, creating what I can only imagine is some Cretaceous version of Robert Smith from The Cure. Mounts like *this* one may do more harm than good in communicating our new image of dinosaurs—a vision in which scaly hides have given way to feathery ones. Like it or not, many dinosaurs were fuzzy, fluffy, and feathery.

Feathers have a very deep evolutionary history. Their trail goes much deeper than the earliest birds, and may even go back as far as the first dinosaurs. Indeed, a flood of fossils discovered over the past fifteen years have irrefutably shown that most, if not all, dinosaur lineages had some kind of feather-like body covering.

The first fluffy dinosaur discovery enthralled paleontologists. At the annual Society of Vertebrate Paleontology conference in 1996, scientists circulated a photograph of a small fossil that revealed a mane of fuzz along a dinosaur's back and tail. John Ostrom, who was chiefly responsible for reinvigorating the idea that birds are dinosaurs, was "in a state of shock" after hearing the

Microraptor is one of more than thirty feathered non-avian dinosaurs found so far. (The white arrows point to feathers on this dinosaur, and the black arrows indicate more subtle feather traces that can be seen only under UV light.) By studying the microscopic structure of *Microraptor* feathers, paleontologists have even discovered that this dinosaur had dark, glossy feathers. In life, it looked something like a toothy raven. (Image from www.plosone.org/article/info%3Adoi%2F10.1371%2Fjournal.pone.0009223)

news. At long last, a feathery non-avian dinosaur had really been found. This creature, labeled *Sinosauropteryx* in a technical publication the same year, didn't have feathers suited for flying. The simple dinofuzz covering the creature's body could only have been for display and insulation—the dinosaur lacked the specialized, asymmetrical feathers that allow modern birds to take to the air. In fact, it would have looked very much like Huxley's hypothetical feathery *Compsognathus*. The newfound dinosaur pointed to the hypothesis that feathers were not originally used for flight, but had evolved for different reasons and were later co-opted.

At least thirty different feathery non-avian dinosaurs have been recognized since that first one. Some are more "birdlike" than others. *Anchiornis*—a roughly 160-million-year-old, pigeon-size dinosaur—had elongated feathers on its arms and legs that might represent an intermediate state between wholly terrestrial dinosaurs and early fliers. And even *Velociraptor*, a turkey-size predator

that most certainly did not fly, had elongated feathers on its arms—
a feature inferred from quill knobs preserved on the dinosaur's
arm bones. If there is ever a *Jurassic Park 4*, and that movie has
Velociraptor reprise its role, the dinosaur should sport some exquisite
plumage, Steven Spielberg's sense of taste be damned.

Even bizarre dinosaurs further removed from the avian rootstock sported decorative, feather-like structures. *Beipiaosaurus
inexpectus*—a potbellied dinosaur with long claws, an extended
neck, and a beaked skull better suited to clipping plants than slicing
flesh—was enveloped in two layers of differentiated, simplified,
elongated feathers. Tyrannosaurs had feathers, too. A small form
named *Dilong* and a much more formidable, 30-foot genus called
Yutyrannus had filamentous coats of fuzz. Thanks to these finds, we
can say that *Tyrannosaurus rex* was probably a feathery giant—an
idea that will undoubtedly cause dinosaur traditionalists to have
a conniption.

Feathers were not just a feature of birds and their closest non-
avian predecessors. Birds are just one lineage of a wider theropod

**Thanks to exquisitely preserved skeletons with intact gut contents, we know that fuzzy
dinosaurs such as *Sinocalliopteryx* fed on their feathery neighbors, including other non-
avian dinosaurs (left) and early birds (right).** (Art by Cheung Chung Tat. Image from www.plosone.org
/article/info%3Adoi%2F10.1371%2Fjournal.pone.0044012)

family called the Coelurosauria. Every lineage within the Coelurosauria has at least one representative with dinofuzz or full-blown feathers. More than that, we now know that feathery adornments were a common dinosaur feature. Two dinosaurs—each about as far removed from birds as possible—also displayed body coverings structurally very similar to simple feathers. *Psittacosaurus*, which looks like an animal with a parrot head and a ceratopsian body, had an array of bristles along its tail. Even though most of its body was covered in scales, the bristles were very similar to the fluffy coatings found on theropod dinosaurs. And another dinosaur named *Tianyulong* sported a row of similar bristly ornaments along its back. These dinosaurs were ornithischians—forms that existed on the other side of the evolutionary tree from the coelurosaurs. Since creatures on both sides of the dinosaur family tree had feathers or feather-like body coverings, the fuzz and bristles might have been a common dinosaur feature, inherited from the last common ancestor of all dinosaurs. And the description of a fuzzy juvenile dinosaur named *Sciurumimus* in 2012—a dinosaur near the base of the theropod family tree, far from birds—added another feathery data point to the idea that protofeathers were

Even though there isn't any direct evidence yet, the discovery that many dinosaurs were partially covered in protofeathers means that some sauropods—such as this juvenile *Apatosaurus*—might have been fuzzy, too. (Art by Niroot Puttapipat)

Even *Tyrannosaurus rex* itself was probably coated in fuzz. Despite complaints from fans of scaly *T. rex*, the carnivore wouldn't have been any less fierce. (Art by Niroot Puttapipat)

widespread among dinosaurs. Most, if not all, dinosaur lineages might have had dinofuzz, and that includes the impressive sauropods. (Just think of how cute a fuzzy little *Apatosaurus* juvenile would be.)

We are left with only two possibilities. Either the same kind of simple filaments evolved over and over again, or dinofuzz was an ancient trait that was present in all dinosaur lineages. I can almost hear the scaly-tyrannosaur fans weeping.

The various types of prehistoric feathers cataloged so far outline how plumage has evolved. As far as paleontologists understand as of this writing, feathers started off as fuzz and in time were adapted into complex structures that allowed some dinosaurs to take to the air. Protofeathers were simple, single filaments. These are the kinds of structures seen on the bodies of *Psittacosaurus*, *Tianyulong*, and *Sciurumimus*. Archaic coelurosaurs—agile little dinosaurs like *Sinosauropteryx*, the first to be recognized with dinofuzz— had slightly more complex coats. Their protofeathers had multiple branches coming out of a central filament. These feathers were not all that different from those seen on oviraptorosaurs—beaked, omnivorous dinosaurs that were already quite birdlike to start with—as well as on parts of some true early birds. (Some feathered dinosaurs and early avians had multiple feather types on their bodies, just like modern birds.)

In the next stage of feather evolution, the individual filaments branched further along a central support. In dinosaurs such as *Microraptor* as well as in the earliest birds themselves, these individual filaments eventually formed true leaf-shaped feathers organized along a central vane. Some of these feathers, like those seen in the flightless raptors, couldn't support dinosaurs in the air, but *flying* dinosaurs—including *Archaeopteryx* and the four-winged *Microraptor*—had specialized, more aerodynamic feathers that were thinner along the leading edge. These were the feathers that finally allowed dinosaurs to invade the skies. Feathers originally formed insulating coats and flashy displays, and at least one lineage co-opted the same structures to become the only flying dinosaurs. Regardless of what *Archaeopteryx* was or was not, the gradual flow of scientific discoveries has revealed dinosaurs as increasingly birdlike and inextricably connected avian dinosaurs to their non-avian forerunners. Fossil feathers solved the mystery.

There's more to the dinosaur-bird connection than avian origins alone. Many of the fantastic new discoveries about dinosaur biology have been influenced by the fact that we have living dinosaurs to study. A chickadee isn't an *Ankylosaurus*, and an emu isn't a *Diplodocus*, but today's birds can help paleontologists refine questions and ideas about how dinosaurs lived. Best of all, our avian dinosaurs can finally help us fill out the palette of their extinct relatives.

As Charles Darwin wrote, "[I]gnorance more frequently begets confidence than does knowledge: it is those who know little, and not those who know much, who so positively assert that this or that problem will never be solved by science." Darwin was referring to humanity's origin—a mystery complicated by elusive evidence and dogmatic religious strictures—yet the same argument applies to the question of dinosaur colors. The problem was not an absolute lack of evidence, but the fact that the stepwise process of scientific understanding has only very recently grasped where to look for the essential clues.

I was reminded of Darwin's line while waiting for a session to start at the Society of Vertebrate Paleontology's 2011 meeting in Las Vegas—quite a setting for a conference on prehistoric life. The constant lights and buzz of Bally's grated on my every nerve, but I tolerated the cigar-smoking gamblers and the bagpiper who played on the street below until the early hours of the morning, because this was the temporary haven for the best and most cutting-edge info on paleontology. I had been waiting all year to hear about the new discoveries being made in the field and the lab. I especially wanted to hear what Brown University graduate student Ryan Carney had to say: he was set to reveal the true color of the first *Archaeopteryx* specimen ever found—the isolated feather used to name the dinosaur 150 years earlier.

The paleoartist Bob Walters sat down to my left a few minutes before the presentation was scheduled to start, notepad at the ready. I jokingly asked if he was angry at paleontologists who stepped on his turf and told him what colors were now considered acceptable. Bob looked shocked. "Not at all!" he said. Artists like Bob had long been hoping for some scientific indication of dinosaur color, he said, and now paleontologists were going to give them just that.

Once Carney took the stage, he didn't waste any time relaying the news of his team's discovery. The *Archaeopteryx* feather was black. Whether the whole animal was black was impossible to say. The single feather was selected because it was a famous specimen and it was the sesquicentennial anniversary of when the beloved feathered dinosaur was named, but nevertheless, the analysis had finally attached a color to one of the world's most important fossils.

The method by which Carney and his collaborators determined the dinosaur's hue was developed several years ago, and it all started with a squid. A very, very old squid, but a squid all the same. Jakob Vinther, a molecular paleobiology graduate student at Yale University, was inspecting the ink sacs of a fossil cephalopod under a high-powered electron microscope when he noticed little blobs inside the membranous pocket. Paleontologists had seen structures like these before and had assumed that they were fossilized bacteria, locked in stone as they started to break down prehistoric soft tissues. But the fact that the microscopic spheres were restricted to the inside of the ink sac suggested something different. These were melanosomes—tiny organelles whose shape, density, and distribution create pigment. In the squid, the melanosomes gave a dark-brown color to the ink the cephalopod used to escape from predators, and Vinther wondered if melanosomes might be detected in other fossils.

Feathers seemed a good place to look because many of their colors are created by melanosomes. If fossil feathers contained melanosomes, and zoologists could examine the feathers of modern birds to see how the organelles corresponded to certain colors, then they could reconstruct the colors of prehistoric creatures. Before Vinther and his collaborators could investigate non-avian dinosaur feathers, they had to establish that they were really seeing melanosomes and not bacteria. They did just that with a fossil feather from Cretaceous Brazil. The feather was banded white and black. If the little round bodies were bacteria, then they should have been found all over the feather's surface. As the researchers discovered, though, the tiny spheres were constrained to the dark sections only. These were the bands that would have carried pigment, and so the scientists could be confident that they had identified real melanosomes.

Vinther knew that the findings had applications for dinosaurs, too. A beaked, birdlike dinosaur named *Caudipteryx*, he and his team pointed out, had a fan of banded tail feathers that might actually represent the true color pattern of the dinosaur. But his study didn't catch the public's attention. This was 2008. The key to dinosaur color had just been found, and yet the implications didn't reach far beyond the small number of researchers who read the paper. Still, the scientists kept at it, and the following year, Vinther led another study on a 47-million-year-old feather found in Germany. This one, from a bird that lived about eighteen million years after the demise of the last non-avian dinosaurs, had an iridescent sheen in life.

Non-avian dinosaurs were next on Vinther's list. But as often happens in paleontology, another team got there first. On January 27, 2010, the Chinese Academy of Sciences paleontologist Zhang Fucheng and a team of collaborators published online a letter in *Nature* about the colors of Cretaceous birds and, for the first time, non-avian dinosaurs. Among other specimens, the team had selected a *Sinosauropteryx*—the fuzzy dinosaur that had marked the onset of a flood of feathered dinosaurs from China starting in 1996. From the time the dinosaur was described, it was apparent that the protofeathers along its tail had a banded pattern. The team took only a very limited sample, but concluded that the darker patches had been reddish-brown. *Sinosauropteryx* had a candy-cane tail that could have been used as a visual signal among these dinosaurs.

A week after the online announcement, Vinther's team countered with an even more detailed study in *Science*. It was the first time a non-avian dinosaur had been fully restored in color. The Beijing Museum of Natural History paleontologist Quanguo Li, Vinther, and collaborators worked with a specimen of *Anchiornis*. This small dinosaur was known from multiple, roughly 160-million-year-old specimens, and it looked something like a magpie. *Anchiornis* was black with swaths of white on its arm and leg feathers. But most impressive of all was a tuft of reddish plumage on the dinosaur's head. I hadn't seen anything like it before. *Anchiornis* looked rather plain, but in a strikingly beautiful way, made all the more wonderful by the fact that we could now tell what color dinosaurs were.

Both *Archaeopteryx* and *Anchiornis* were at least partly covered in black feathers. They looked like modern crows rather than birds of paradise. Vinther and colleagues found similar hues when they looked at the feathered dinosaur *Microraptor*. Using the same techniques on an absolutely gorgeous specimen of this sickle-clawed dinosaur, the team discovered that it boasted a glossy coat of complex feathers. Like *Anchiornis* and *Archaeopteryx*, *Microraptor* was a dark-colored dinosaur which wouldn't look out of place perching with the ravens along a western highway.

It never ceases to amaze me that we can now tease out dinosaur colors from the fossil record. The implications go far beyond the artist's palette choices. Whether stripes, spots, or iridescent plumage, feathered dinosaurs boasted visually arresting patterns. These dinosaurs were highly visual creatures who communicated with lovely, colorful displays. Even better, as we study more specimens from each species, we'll be able to investigate whether dinosaurs had different color patterns in each sex or distinct breeding plumage. Color might be the key to other aspects of dinosaur biology.

So far, the technique works only for dinosaurs with preserved feathers. For species of dinosaur that didn't have feathers, or even specimens of feathered species that were preserved without their plumage, we can't investigate their colors. We're also still looking for a way to detect and restore chemically created colors—some of the greens, blues, oranges, and yellows seen in many birds. As far as the science can reach, at least at the moment, there needs to be something to preserve the melanosomes and to be compared to modern analogs. You can't draw blood from stone, but if you know how to look, you can get dinosaur colors.

We're fleshing out the old bones that pack museums around

the globe. We've uncovered intricately preserved specimens, reconstructed their body coverings, and now have a good sense of what they looked like. So with all of this in mind, how did dinosaurs see each other? Birds can see ultraviolet parts of the spectrum—could *Microraptor* have exchanged visual cues that we could never hope to see? What if there were a way to get inside a dinosaur's head, see the world through her eyes, and understand how she perceived her surroundings?

Eight

Hadrosaur Harmonics
and Tyrannosaur Tastes

What is a dinosaur without a roar? From awkward, jerking robots in traveling museum exhibits to the best special effects Hollywood can muster, the ability to thunder an ear-shattering, gut-shaking primal scream is what brings dinosaurs to life. *Unknown Island*—the 1948 stinker that was the first to show rampaging dinosaurs in color—wouldn't have been the same without the chaotic cacophony of *Ceratosaurus* screeches, and *Jurassic Park* wouldn't have seemed so dangerous without the terrifying bluster of *Tyrannosaurus*. And it's not just the carnivores. Imposing herbivores like *Triceratops*, *Brachiosaurus*, and *Ankylosaurus* come to life through their hoots, snorts, and snuffles, too. Even in *Discovery*'s docudrama *Dinosaur Revolution*, dinosaurs are all about grunting and snarling their archosaurian hearts out. If cinema and basic-cable documentaries are to be believed, dinosaurs were some of the chattiest creatures of all time.

Like many intricate details of dinosaur lives we see on screen, these distinctive sounds are speculative. Dinosaur noises in pop culture are often composed from various animal sounds spun together in a blender. The bone-rattling roar of *Tyrannosaurus* in *Jurassic Park* is actually a combination of elephant, alligator, tiger, dog, and penguin sounds. I learned early on in elementary school art class that mixing enough colors made brown, and it would seem that if you blend enough animal noises together you ultimately get a dinosaur bellow.

Far too much time spent in front of flickering screens taught me that dinosaurs must have had impressive voices to match their imposing frames, but I have always wondered what they really sounded like. The Fayetteville State University paleontologist Phil Senter has pondered the same question in his paper "Voices of the Past: A Review of Paleozoic and Mesozoic Animal Sounds." I hoped Senter had had access to both prehistoric dinosaurs and a high-fidelity recording device. Sadly, this wasn't the case, and as he reminds readers, "The fossil record does not include audio recordings." Damn.

Lacking direct access to dinosaur sounds, Senter resorted to the tried-and-true method of using modern animals to investigate what prehistoric creatures might have sounded like. In the case of the dinosaurian multitude, of course, this meant turning to crocodylians and avian dinosaurs for clues. The trouble is that the two archosaur groups don't produce sound in the same way. While crocodylians vocalize through their larynx, birds sing and chirp via a different organ, called the syrinx—a series of cartilage rings deeper down in the upper respiratory tract. Since these structures are so different, Senter argued, the sound-producing abilities of crocs and birds must have evolved independently of each other. Therefore, Senter concluded, dinosaurs probably didn't scream, sing, or roar. As a consolation, he proposed that dinosaurs made sounds by other means—hissing, clapping their jaws together, rubbing their scales, splish-splashing in the water, and, for *Apatosaurus* and similarly equipped sauropods, cracking their whiplike tails. That's not especially satisfying to someone raised on especially vocal *Tyrannosaurus* and *Triceratops*.

I hadn't considered all the different ways dinosaurs could have made sounds. When rival *Spinosaurus* met on ancient floodplains, did they clap their crocodile-like jaws to threaten each other? Did a mother *Oviraptor* hiss at intruders who ambled too close to her nest? Scenarios like these are certainly possible, but I still believe that dinosaurs had voices, too. After all, even though living dinosaurs—birds—and the closest living cousins of dinosaurs—crocodylians—vocalize differently, their common ancestor wasn't *necessarily* mute. Birds still have a larynx, even though they don't use it to make sounds, and crocodylians are vocal enough via the same structure. Perhaps most non-avian dinosaurs had a more croc-like mode of expressing themselves before the specializations that allow birds to sing evolved. When I visited some baby alligators in their rooftop greenhouse at the University of Utah, I mostly heard hisses—the sharp-toothed babies were not very happy that I had stopped in. But take just a few minutes with YouTube and you'll find that you can sample a few other crocodylian noises too. Take the short clip of a male American alligator at the St. Augustine Alligator Farm Zoological Park. The alligator's rumble sounds like a cross between a clogged drain that's suddenly unstuck and a failed attempt to start a lawn mower. Another clip, filmed in the Everglades National Park, shows a pair of males—heads raised, tails lifted high out of the water—making similar noises in territorial displays. The water dances over their armor-covered backs as they call to each other.

Dinosaurs may have heard the prehistoric relatives of alligators and crocodiles make noises like this in prehistoric swamps—just imagine the sounds a forty-foot alligatoroid like *Deinosuchus* would have been capable of—but whether they made similar sounds is speculative at best. Tempting as it is, we can't really take a handful of birds and crocodylians and extrapolate the sounds made by those animals to a vast range of different dinosaurs. That alligators and crocodiles vocalize is a clue that dinosaurs could have called to each other, but we can only try to reconstruct dinosaur sounds from the little evidence we have.

Obviously, we would need to know a lot more about dinosaur soft tissues—the anatomy of their throats, especially—to reconstruct the sounds they made. And, fortuitously, there are some delicate skeletal clues that can help us determine how dinosaurs made themselves heard. At least one group of dinosaurs exhibit clues to their sonic abilities in their elegant headgear. *Parasaurolophus* was one such Mesozoic musician. From neck to tail, this hadrosaur wasn't that different from its close relatives. The body of this dinosaur is just about as average as you can get—*Parasaurolophus* had the short arms, long legs, and deep tail that were typical of hadrosaurs. What immediately sets *Parasaurolophus* apart is its gorgeous skull. Jutting out from the back of the dinosaur's cranium is a slightly curved, tubular crest. The development of the crest differs from one species to the next. *Parasaurolophus cyrtocristatus* from Late Cretaceous New Mexico and Utah had a shorter crest than its slightly later relatives *P. tubicen* from New Mexico and *P. walkeri* from Alberta, Canada. Nevertheless, all three share similar, instantly recognizable adornments. The beauty of the *Parasaurolophus* ornamentation isn't just the adornment's outward appearance. Having a big tube sticking out of your skull is one way to get attention, but it's the internal anatomy of the dinosaur's crest that held the real communication secret.

I was fortunate enough to see the feature firsthand when I visited one of Utah's vast southern badlands on a sunny May morning in 2010. Grand Staircase–Escalante National Monument is a gorgeous, isolated swath of southern Utah desert. During my trip to the park's Cretaceous exposures, the paved road rapidly gave way to a well-groomed dirt track that changed colors with the rocky, sage-filled landscape on the route to the pastel yellow monument dubbed Grosvenor Arch. I wished I had the chance to enjoy the rest of the park's scenery beyond that point. The flattened trail quickly turned to a long one-lane runnel studded with rocks and pocked by potholes as the track—a road in name only—oscillated up and down on the way up to the Kaiparowits Plateau.

I tried to calm myself as my car rattled and jolted over "The Cockscomb" and down "The Gut." I had spent my entire life in the well-paved eastern states, and the roughest roads I had typically encountered were gravel parking lots. I kept a death grip on the steering wheel, stared straight ahead, and tried not to imagine

The hollow crests of some hadrosaurs—such as this *Parasaurolophus*— may have allowed these dinosaurs to bellow a range of low notes over long distances. (Photograph by the author at the Natural History Museum of Utah)

the conversations with my insurance agent about what had happened to the rental car I was driving. ("Now, what were you doing
at the time of the incident?") The last stretch before I hit the University of Utah camp on Horse Mountain was the worst—the last
jog of dirt road looked like something out of an Indiana Jones
movie. A few small chunks of road had fallen away along the
curve, a definite reminder to aim very carefully. Imagine my de-
light when, the morning after my arrival at camp, the experienced fossil finder Scott Richardson had me take the same route
in the other direction to get down to a very special fossil out in the
monument.

The exposures in the BLM-managed park aren't quite what
you'd expect for the West. Parts of the Kaiparowits Formation—a
roughly 75-million-year-old stack of rocks formed when southern
Utah was a coastal swamp near a shallow seaway that cleft North
America in two—are covered by shrubs and short trees sunk into
sandy soil. The dinosaurs in these shrub-covered sections of the
formation don't so much tumble out of naked rock as they seem to
roll out of the dirt. After Scott and I parked our cars and started
out to where the *Parasaurolophus* was, Richardson pointed out a
solitary dinosaur toe bone resting in the soil. The chocolate-colored
bone had simply fallen out of the encasing sediment, though
Richardson couldn't touch it. There are strict and specific BLM
rules about collecting fossils in this area, and the proper permits
had to be arranged before plucking up so much as a fragment of
bone. The dinosaur toe would have to wait.

When we finally arrived where the *Parasaurolophus* rested, I
didn't immediately understand what I was looking at. All I could
see was a strange dark patch on a huge block of gray rock. The
exposed part of the fossil looked a little bit like the cartoon illustrations of chromosomes I used to cut out for oversimplified class
projects in Biology 101: inside a rind of tan bone were two pairs of
oblong blobs, the bottom blobs longer than the top ones. I suddenly
realized I had seen this dinosaur shape before, and Richardson

confirmed my guess. I was looking at a cross section through the crest of *Parasaurolophus*, looking into the hollow tubes that ran through the skull ornament.

As I found out when I later looked up the history of this hadrosaur, the Canadian paleontologist William Parks saw something similar when he described *Parasaurolophus walkeri* in 1922. While the dinosaur's elongated crest looked solid enough on the outside, a fortuitous break showed that the structure was actually hollow. Thin partitions of bone separated a pair of tubes that ran from the nose to the back of the skull and down into the mouth—a super-long nasal passage unlike any seen before.

Why a dinosaur would need such an elaborate set of skull plumbing was a mystery. Even more perplexing was the fact that closely related hadrosaurs had similar arrangements, but contained within different crest shapes. In the greater hadrosaur group, there are three main branches: the hadrosaurines, the saurolophines, and the lambeosaurines. The first group, which includes dinosaurs such as *Hadrosaurus* from my former home state of New Jersey, were unornamented forms; and the saurolophines—such as the seemingly ubiquitous *Edmontosaurus* and the "caring mother" dinosaur *Maiasaura*—either lacked flashy ornaments or had relatively simply bumps and crests on their heads. The lambeosaurines had more charismatic headgear. While similar in body to their comparatively plain relatives, these dinosaurs typically had elaborate crests of various shapes. In addition to the tube-headed *Parasaurolophus*, other classic members of this group include *Corythosaurus* (a dinosaur with a domed crest) and *Lambeosaurus* (a form with a large L-shaped crest). In each genus and species, the nasal passage invades the crest to create a circuitous pathway from nose to mouth.

When I was a kid, the outdated books in my school library and a cheap dinosaur album my parents bought me at a local Hallmark store were unanimous about the crest's function. The ornate headgear was obviously used as a dinosaurian aqualung that allowed *Parasaurolophus* and other lambeosaurines to spend most of their time underwater. In his children's book *Dinosaurs and Other Prehistoric Reptiles*, artist Rudolph Zallinger painted a *Parasaurolophus* and *Corythosaurus* synchronized-swimming under the surface of a Cretaceous lake, and a sticker book created by the Italian company Panini depicted a whole herd of *Parasaurolophus* grazing underwater. This made perfect sense within the completely wrong supposition that these dinosaurs were amphibious herbivores. After all, the hadrosaurs had been named "duck-billed" dinosaurs—a name that has stuck despite its inaccuracy. Well-preserved hadrosaur specimens have shown that these dinosaurs had expanded beaks that looked like a pair of interlocking vertical shovels. And certain hadrosaur specimens with intact skin impressions were misinterpreted to cast the dinosaurs as dedicated swamp dwellers. In 1912, the paleontologist Henry Fairfield Osborn described an exceptional *Edmontosaurus* specimen that had been discovered by Charles H. Sternberg, with preserved skin impressions over most of its body. It appeared as though the dinosaur's hand was enclosed within a fleshy mitt. Osborn believed that this was a kind of webbing that would have helped the dinosaur scull around Cretaceous lakes and rivers. Actually, the condition was a common hadrosaur feature that helped keep the "fingers" together in a single unit for support while walking on all fours.

Before we knew the truth about hadrosaurs, an aquatic mode of life seemed like the perfect explanation for the weird crests of some species. While Parks was uncertain about the function of the *Parasaurolophus* crest—maybe the feature acted as some kind of visual signal?—in 1933 Alfred Sherwood Romer suggested two possible hypotheses. Maybe the crests acted as air tanks, or perhaps they allowed the dinosaurs to snorkel. The fact that there were no external crest openings for the dinosaurs to breathe through sank the snorkel idea, but, as I would learn so many years later, the air tank idea had popular appeal.

Not that all paleontologists agreed with Romer's notion, but other experts still envisioned the crests as having something to do with an aquatic setting. Charles Mortram Sternberg—one of the sons of the famous fossil collector Charles H.—suspected that the U-shaped bend in what he called "hooded hadrosaurs" prevented water from entering the respiratory system of dinosaurs like *Lamebosaurus* when they dunked their heads underwater to feed. The German paleontologist Martin Wilfarth had a more fanciful idea. The expanded areas of bone were anchors for trunks, he proposed, which would have allowed *Corythosaurus* and kin to reach up to the surface for a breath of air when needed.

Even John Ostrom, who criticized and tossed out the other ideas, still viewed the expanded nasal passages within a watery milieu. "It is quite probable that hadrosaurs lived a rather passive, perhaps even retiring existence," he wrote, and pointed out that "[t]hey possessed no horns, no claws, no sharp teeth, they carried no clubbed or spiked tail, and they had no bony armor. They certainly were not constructed for rapid flight and they cannot be considered giants for their time." What was a defenseless hadrosaur to do?

Ostrom thought their nasal extensions might be the solution. The passageways could have increased their ability to smell predators from a long way off and escape into the water when necessary. This idea didn't catch on either. There was no indication that the part of the hadrosaur nose that detected smells stretched much beyond the area where the nostrils met the outside world; furthermore, the need to smell predators would predict that hadrosaur crests would have a consistent shape that would optimize the dinosaurs' sniffing ability. The fact that hadrosaur crests were so diverse hinted at a different explanation.

As it turned out, Parks had been on the right track. When James Hopson reviewed the various theories about hadrosaur crests in 1975, he concluded that the structures were visual signals. The main purpose of a crest was simply to be seen. But Hopson cautioned that this scenario did not preclude other explanations.

The long, circuitous passageways could have acted as resonating chambers that would have allowed dinosaurs like *Parasaurolophus* to bellow to each other over long distances. Years before, when the Swedish paleontologist Carl Wiman described the species *Parasaurolophus tubicen*, he remarked that the dinosaur's crest resembled a musical instrument called a crumhorn. I truly hope *Parasaurolophus* sounded more impressive than the woodwind instrument, which reminds me of a fancy kazoo. If *Parasaurolophus*, a dinosaur more than thirty feet long that weighed nearly three tons, sounded so wimpy, it must have been one of the most unintentionally hilarious creatures of all time.

Fortunately, thanks to researchers who have investigated the acoustic abilities of *Parasaurolophus*, we know that the dinosaurs didn't sound so ridiculous. The Johns Hopkins University paleontologist David Weishampel used an improvised model of a *Parasaurolophus* crest to explore the dinosaur's range. I remember seeing Weishampel toot his dinosaurian horn in more than one documentary when I was a kid. Painted with green and orange stripes, the instrument was simply a U-shaped horn made out of PVC pipe roughly the length of a *Parasaurolophus* crest. The sound was like a foghorn—a deep, booming call that would have carried far over the Cretaceous bayous where this dinosaur lived. Watching the clips over again now, the demonstration reminds me of one of Ray Bradbury's short stories. In "The Foghorn," Bradbury imagined a lonely prehistoric monster rising from the sea upon hearing a foghorn—a sound it had mistaken for the voice of one of its own. Alas, the sound of Weishampel's hadrosaur tuba has never drawn out any hiding dinosaurs.

Of course, dinosaurs weren't creatures of paint and PVC. Hadrosaur crests were complex structures of bone and soft tissues connected to the biological rhythms of a living animal. The models can be only vague approximations of how dinosaurs sounded. But there are subtle clues that hadrosaurs made such calls, and that their repertoire even changed as they grew. The evidence is in their ears.

Weishampel got some documentary airtime thanks to his own instrument, but he also wrote several papers on the acoustic properties of hadrosaur crests. From a musical perspective, he determined, the well-ornamented *Parausaurolophus walkeri* had a range from G two octaves below middle C to B below middle C, while the blunter-crested *Parausarolophus cyrtocristatus* was capable of sounds from D one octave below middle C to F# above middle C. And, as Weishamphel outlined in a 1981 *Paleobiology* paper, the delicate ear anatomy of hadrosaurs such as *Corythosaurus* seemed to be consistent with the idea that these dinosaurs were sensitive to a wide range of low-frequency sounds of the sort their crests could have made.

This applied only to the adult animals. Juvenile hadrosaurs, Weishampel hypothesized, lacked the well-developed crests of the adults and may have chirped like baby alligators do. There was a good reason for this. The calls of the little dinosaurs were high-pitched so that their squeaks didn't travel too far and garner unwanted attention from predators. As the dinosaurs grew, however, communication with other members of the same species particularly potential mates—became especially important, and talking long distance is best accomplished through low-frequency sounds. Modern African elephants are a rough parallel here. The rumbles of these mammals allow the modern behemoths to chat from afar.

Weishampel based his hypothesis on previous studies of hadrosaur ears and the anatomy of modern birds and crocodiles. For the most part, paleontologists didn't have much to go on when it came to dinosaur ears. Well-preserved hadrosaur skulls were rare and valuable, so destroying part of a skull to look inside the ear was not acceptable; and while paleontologists had begun employing CT scanners that allowed a look inside at dinosaur skulls, they were expensive to use. It has been only recently, as we develop more accessible and higher-resolution scanning technologies, that we've been able to tell what hadrosaurs were actually capable of hearing.

David Evans is one of the paleontologists who picked up Weishampel's work on hadrosaur sound. In a 2006 *Paleobiology* study he concluded that display, sound, and possible physiological benefits shaped different hadrosaur crests, and in 2009 Evans teamed up with colleagues Ryan Ridgely and Lawrence Witmer to investigate how the development of the dinosaurs' crests, nasal passages, and brains related to each other. After creating three-dimensional scans of skulls representing different age stages of *Lambeosaurus*, *Corythosaurus*, and another dome-crested form called *Hypacrosaurus*, the paleontologists tracked the nasal passages and investigated the anatomy of the brain and inner ear.

The skulls of these dinosaurs held some surprises. In a speci-men of *Hypacrosaurus altispinus*, the researchers found that the di-nosaur's nasal passage looped around in a very elaborate path. As air entered the dinosaur's nose, it traveled up to about the level of the eyes before dropping back down, twisting, rising back up, peaking high in the crest, and looping back downward into the throat. In a juvenile *Lambeosaurus*, by contrast, the nasal passage followed a simpler path in which it swerved briefly at a portion called the S-loop before rising up into the young dinosaur's under-developed crest and coming back down into the throat. In *Hypacro-saurus*, the nasal passage was far more circuitous than was apparent from the dinosaur's rather plain crest. Looking at the outside of the skull, you wouldn't guess that it contained such complicated plumbing. Whatever function the nasal passages had, they must have evolved for different reasons than the crest. The outside form of the crest was probably modified as a biological signpost, and the internal nasal passages were adapted according to different pres-sures relating to sound and physiology.

We can deduce why a dinosaur might have such complicated nasal plumbing by looking inside its braincase. Paleontologists can create detailed virtual models of dinosaur brains by scanning the vacant brain cavities inside fossil skulls. The various lumps and bumps outline how important certain senses were to a dinosaur. Likewise, high-definition scans of dinosaurs' inner ears can help us understand what these dinosaurs could actually hear.

When Evans, Ridgely, and Witmer looked at the inner ears of the hadrosaurs, they found that juvenile dinosaurs were better attuned to picking up a wider range of sound, while adults specialized in detecting lower frequencies of the sort their crests were capable of making. More than that, the brains of these dinosaurs had large cerebral hemispheres—the part of the brain associated with complex behaviors. While not absolutely definitive proof that creating and hearing sounds drove the evolution of hadrosaur headgear, the conclusion was consistent with the idea that dinosaurs like *Parasaurolophus* were vocal, social creatures who communicated with each other from the time they hatched to the time they died.

The same principle can help us outline what other dinosaurs might have sounded like. Most dinosaurs didn't have musical crests, but we can gain some insight into their vocal range by estimating what particular species could hear.

In 2005, Otto Gleich and colleagues used the relationship between inner-ear anatomy and the ability to detect sound in birds and crocodylians to estimate what the huge dinosaurs *Allosaurus* and *Giraffatitan* could hear. In their estimation, dinosaurs were attuned to very low frequencies—lower than 3 kiloherz. As journalists were quick to point out when Gleich's study was published, 3 kHz is about the pitch of a human scream, so should you ever

Should you ever find yourself in the company of big predatory dinosaurs, remember this—they probably won't be able to hear you. (Cartoon by Mike Keesey)

find yourself in the Jurassic, you can shriek to your heart's content without having to worry about the biggest dinosaurs homing in on you. But, barring such time slips, the important lesson is that dinosaur ears may be the best way for us to understand what sorts of sounds dinosaurs made.

Brain scans can do far more than give us a rough estimate of what dinosaurs heard. Reconstructions and casts of dinosaur brains also contain information about their other senses, such as eyesight and smell. Lawrence Witmer has been at the forefront of these investigations. While Witmer spends much of his time considering the sensory lives of prehistoric animals, he has tackled these questions through an approach melding old-school comparative anatomy with high-tech innovations. By understanding how certain structures function in modern animals, Witmer's lab has begun to reconstruct aspects of dinosaur lives that were once thought to be beyond the realm of scientific inquiry.

In late 2011, Witmer, Darla Zelenitsky, and others used reconstructed brain images to track how dinosaurs' sense of smell changed through time. The key was the olfactory lobes. These are the parts of the brain that process incoming information about smells, and the general rule goes, the larger the olfactory lobe, the better the animal can sniff out odors. The idea has to do with what's called the principle of proper mass: the bigger a portion of the brain is relative to other regions, the more important that particular function is to the animal's life. In an animal that primarily relies on sight, you would expect the parts of the brain that process visual information to be relatively large. The same is true with smell.

Some dinosaurs turned out to have very fine-tuned senses of smell. *Bambiraptor*—a switchblade-clawed predator not as cuddly as the name suggests—had a brain about as dedicated to smell as a turkey vulture or a black-footed albatross. Both of these are carnivorous birds that pick up on scents to find food, and it's not a stretch to think that feathery little *Bambiraptor* did the same. Looking at the big picture of bird evolution, the avian descendants of the deinonychosaurs—the kin of *Bambiraptor*—retained their good scenting abilities during their early evolution. It was only later, as birds continued to diversify, that some lineages became more visually oriented while other groups retained (or re-evolved) a powerful sense of smell.

Raptors weren't the only dinosaurs with especially sensitive noses. *Tyrannosaurus rex* is notorious for having very large olfactory bulbs. This feature, along with several of the dinosaur's other peculiarities, put it at the center of a strange debate that has had more to do with Cretaceous PR than with science.

For as long as I can remember, the predatory power of *Tyrannosaurus* was self-evident. The answer to the question "What did a *T. rex* eat?" was "Anything it damn well pleased." But early on, some paleontologists saw tyrannosaurs as immense, clumsy scavengers. In 1917, nine years after *Tyrannosaurus* itself was named, the Canadian paleontologist Lawrence Lambe proposed that the large, slender tyrannosaur *Gorgosaurus* was a "recumbent" carnivore and "played its useful part in nature" as a living garbage disposal—cleaning up the carcasses of ceratopsids and hadrosaurs from the Cretaceous landscape.

Decades later, Jack Horner proposed a scavenging life for *Tyrannosaurus rex*, too. At 1994's Dino Fest conference, Horner pointed out that *T. rex* had "beady eyes," tiny arms, and a deep skull better suited to crushing than slicing. The tyrant seemed less a predator than a scavenger, and this controversial interpretation instantly became a news hook. Nearly every *Tyrannosaurus* discovery—about how fast the dinosaur could run or how powerfully it could bite, based on tooth-marked remains of other dinosaurs it had consumed—was brought into the context of whether the dinosaur was a consummate predator or a filthy scavenger.

Most paleontologists didn't see the point of the controversy.

Tyrannosaurus was certainly capable of both hunting and scavenging. The tyrant did have a fine-tuned sense of smell, but the carnivore also had forward-oriented eyes—*T. rex* was one of the few dinosaurs that could have zeroed in on you with binocular vision. And even though the dinosaur's arms have frequently been ridiculed, *T. rex* traded grasping arms for a heavy skull capable of delivering devastating bites. There's nothing about the dinosaur that would have precluded the tyrant from hunting *Edmontosaurus* and *Triceratops*, or from wolfing down carrion, when the opportunities presented themselves. *T. rex* was an all-purpose carnivore, capable of dismantling rotting carcasses as well as taking down prey on the hoof.

The intricate details of the tyrant's skull can give us an idea of how the dinosaur perceived the outside world. Thanks to new techniques and a renewed interest in dinosaurs as animals, paleontologists are slowly putting together a more comprehensive understanding of how dinosaurs interacted with the world around them. Oddly enough, some of the most detailed information about dinosaur biology doesn't come from imagining how they walked, bit, and fought, but from signs of disease and injury preserved in their skeletons. The ailments that chipped away at a dinosaur's health can tell stories about how the animal actually lived.

Nine

In the Bones

Tyrannosaurus rex is immortal. In fiction and in scientific restoration, no creature has ever been as vicious or fearsome. The dinosaur's name isn't just a figurehead title, but a reminder that flesh-tearing, bone-crushing power reached its apex in *Tyrannosaurus*. In fiction, especially, *Tyrannosaurus* is less an animal than a force of nature. The dinosaur's steps shook the earth, its hellish howl had the power of a hurricane, and *Tyrannosaurus* was beautifully adapted through millions of years of evolution to efficiently separate flesh from bone. *Tyrannosaurus* is the epitome of all that is truly wild. Maybe that's why we refuse to believe that this dinosaur had more than a few weak spots. Indeed, until I started riffling through the literature on paleopathology, I had no idea that dinosaurs frequently suffered broken bones, scratched at parasites, fought off infection, and otherwise faced the slings and arrows of life.

Sue suffered plenty. She is the most complete *Tyrannosaurus* ever found, as well as one of the largest, and is a fossilized lesson in Cretaceous hardship. Her battered skeleton is on display at Chicago's Field Museum. The last time I came close to her digs, I was moving cross-country to my new home in Utah with three pissed-off cats in the backseat of my tiny car. A brief stop at the Field was out of the question, but fortunately for me, I knew that a Sue facsimile was going to be reassembled in my neighborhood soon. I'd at least get the chance to visit a Sue clone. The Museum of Idaho, part of the urban-suburban sprawl of Idaho Falls a few hours north from Salt Lake City, was set to host a traveling exhibit about the world's most famous *Tyrannosaurus rex*.

I did a little homework before I drove up to Idaho. There were specific parts of Sue I wanted to check out. I had heard that she'd endured several broken bones, and that an unknown trauma had injured her jaws. I wanted to be able to focus on these damaged bones. It's easy to look at a dinosaur skeleton without really seeing it, the same way you might look at a house, office, or church and overlook the small details. There are subtle clues hidden in the awesome architecture of dinosaur skeletons—bumps, knobs, muscle scars, and flanges that together represent the creature's essential nature and tell us about its life. These are the signposts declaring that we're not just looking at static lumps of rock, but at the remains of a once-living animal more fantastic than anything we could have imagined on our own.

I studied the University of Iowa paleontologist Chris Brochu's exhaustive monograph on Sue's skeleton. Brochu found multiple signs that the *Tyrannosaurus* had seen some combat in her life. Several of Sue's ribs were broken, as well as the dinosaur's right humerus and shoulder. The damage on her right side was probably caused by a single traumatic event, he reasoned, and fractured ribs on the dinosaur's left side, too, hinted that broken bones were just part of life for a successful tyrannosaur. And internal infections had altered the bones in other parts of Sue's skeleton, such as the left fibula and two of the vertebrae. Sue survived: the tyrant's bones show signs of healing, and she didn't ultimately collapse from an infection from the breaks. But Brochu noticed something critical when he began to explore the architecture of Sue's jaws. He saw that they showed signs of extensive damage that may have made it impossible for her to eat toward the end of her life. This conundrum especially intrigued me—what could have weakened the massive tyrant?

When I arrived at the Idaho museum, I walked right past the small gallery of hands-on exhibits and headed straight for Sue. What I wanted to know wasn't in the traveling exhibit's signage, but in the dinosaur's bones. Fortunately for me, Sue was big enough that—even behind the exhibit barrier—the tyrannosaur's trauma is easy to spot. Gnarly lumps of bone gave away healed breaks and places where the dinosaur's body fought off infection, but most impressive of all were those jaw wounds. Sue's lower jaws look as if someone unloaded a bevy of heavy-gauge shotgun shells into the dinosaur's mouth. Both sides of her jaws are perforated with large, smooth holes.

Shortly after Sue was discovered, Peter Larson—part of the crew that unearthed the tyrannosaur—said that the holes were bite wounds. This hypothesis fit a pattern of injury other paleontologists had found among other tyrannosaurs. In a survey of theropod head wounds, Darren Tanke and Philip Currie discovered that large predatory dinosaurs often fought by biting each other on the face. Particular specimens of the tyrannosaurs *Albertosaurus*, *Gorgosaurus*, and *Daspletosaurus* all had bite wounds on their skulls that could have been inflicted only by their own kind. It wasn't a stretch to think that *Tyrannosaurus* did the same, and the fact that the holes in Sue's lower jaw didn't show any sign of healing meant that this mighty dinosaur was attacked and killed by a rival.

But something was amiss. As Brochu and other paleontologists examined the jaw injuries, they realized that the distribution and shape of the holes didn't match a tyrannosaur's teeth. The pathology wasn't what scientists would expect for a crushing bite from another huge *Tyrannosaurus*. The question of who killed Sue was reopened, and paleontologists now suspect that organisms much, much smaller than any dinosaur were to blame.

The veterinarian Ewan Wolff, the paleontologist Steven Salis-
bury, and their team solved the mystery of Sue's demise in 2009.
They announced that Sue and her fellow tyrannosaurs were afflicted by a microorganism that commonly infests the mouths and
throats of modern hawks. In living birds, the microscopic pest is
called *Trichomonas gallinae*, and the mode of transmission is simple:
pigeons (for example) drink water that harbors the microorganism, the parasite takes up residence in the birds and infects them,
and raptors that prey on pigeons get infested too. There are various strains, and some of them show no symptoms, but others eat
away at the lower jaw of the bird, creating lesions in the bone and
ulcers in the soft tissue. In the most severe cases, the damage is so
severe the birds can no longer eat or drink properly.

While the exact same species of parasite wasn't around when
Tyrannosaurus lived, Sue and other tyrannosaurs were undoubtedly

**At the end of her life, *Tyrannosaurus* Sue suffered from an oral infestation of parasites
that ate away her jawbones to the point where she may not have been able to eat.** (Illustration by Chris Glen, The University of Queensland, from doi:10.1371/journal.pone.0007288.g004)

afflicted by a hitchhiker of the same ilk. (Indeed, if you look closely at many *Tyrannosaurus* skeletons on display at museums around the world, you can easily spot these large, smooth-sided holes.) Sue surely suffered. The bone lesions, dead tissue, and ulcers must have made her normally enjoyable feasts (or so I'd like to imagine) so painful and difficult that Sue simply starved to death. The most fearsome carnivore of all time was brought to its knees by an even more vicious predator so small that it can't even be seen with the naked eye.

No one knows how Sue wound up with a mouthful of harmful microorganisms. There's more than one possibility. Thanks to the specimens Currie and Tanke described, as well as punctures on the snout of a young *T. rex* nicknamed "Jane," we know that tyrannosaurs bit each other on the face when they tussled. And, if the attacking tyrannosaur had a mouth brimming with parasites, this would be a great way to transmit the disease. Coarsely serrated teeth were comfortable homes for bacteria and protozoa feeding on tidbits of leftover meat from dinosaur dinners, and such dirty teeth would have driven deadly microorganisms deep into the heads of other tyrannosaurs. Yet Sue doesn't have any bite marks. If an infested tyrannosaur didn't attack her, then the parasite must have found its way into her system some other way.

Cannibalism is another possibility. Of the many *T. rex* specimens collected to date, at least four are gouged by bite marks made by another large, carnivorous dinosaur. The details of these fossils show that the tooth marks were made after death and record feeding, rather than fighting. In the Hell Creek Formation in Montana where we've found *Tyrannosaurus* remains, there has been only one predator of the size and power to inflict such damage—and that's *T. rex* itself. *Tyrannosaurus* ate their own kind when they had the chance. And if a *Tyrannosaurus* just so happened to feed upon one of its kin that was infested with parasites, maybe the little buggers could have jumped ship. Cannibalism is a great way to spread diseases around.

Sue's perforated jaws reveal that other organisms harried, invaded, and infested even the mightiest dinosaurs. Their damage can't always be seen on the bones; Sue's injured jaws were an exceptional case. For the most part, parasites thrived in and on the soft tissues of dinosaurs. When their host's body decayed, the parasites disappeared, too. But a few unusual fossils show that dinosaurs provided homes for whole ecosystems of parasites.

Terrible prehistoric parasites bring to mind a modified version of the tagline for John Carpenter's body-snatching gore fest *The Thing*: for some parasites, dinosaurs were the warmest places to hide. And since dinosaurs were so fantastic, we might expect their parasites to be even more horrifying. In the realm of fiction, we've had some fun conjuring up how awful dinosaur parasites must have been. When he created a lush jungle filled with prehistoric survivors for his *King Kong* remake, director Peter Jackson conjured the imaginary carrion-eating monster "*Carnictis*"—a kind of flesh-eating worm that lived in the guts of tyrannosaurs, but took on a gruesome existence of its own when a dead dinosaur's viscera spilled out into a rocky pool of just the right conditions. Long before that, in his short story "Poor Little Warrior!," science fiction author Brian Aldiss imagined the horrible little arthropods that swarmed over the bodies of the biggest dinosaurs. When Claude Ford, a time-traveling hunter bent on proving his masculinity by gunning down a monstrous trophy, fells a wallowing sauropod, the creature's battalions of parasites look for the closest warm body they can find. That happens to be Claude. "You struggle and scream as lobster claws tear at your neck and throat. You try to pick up your rifle but cannot, so in agony you roll over, and next second the crab-thing is greedying it on your chest," Aldiss imagined. The thought had never occurred to a "little shrimp" like Claude that a dinosaur's parasite "would be a good deal more dangerous than their host." (And in the realm of science *fact*, in *Parasite Rex*, the science writer Carl Zimmer wondered if dinosaurs hosted tapeworms, and if so, how terrifying *those* parasites must have been. Big dinosaurs could have sustained gargantuan parasites. Thus far, no direct evidence of tyrannosaur tapeworms had ever been found, but it's not out of the question.)

Sadly for B-movie fans, most dinosaur parasites were not actually as gigantic or as terrifying as we have feared. They weren't so different from modern parasites. We know this, at least in part, thanks to dinosaur shit. Much like fossil tracks, dinosaur coprolites—or fossilized feces—don't get the attention they deserve. You're never going to see a blockbuster museum exhibit centered around the great piles deposited onto Mesozoic soil. But preserved dinosaur scat contains plenty of information worth sharing about dinosaur biology. Tyrannosaur coprolites have shown that these dinosaurs swallowed huge quantities of flesh and bone, and their digestive systems were so fast that they didn't even completely digest the parts of other dinosaurs they ate. And the pats left behind by sauropod dinosaurs have helped paleontologists track the makeup of prehistoric ecologies, as well as the evolution of grasses and other plants. Furthermore, some organisms considered dinosaur crap to be both food and shelter—some well-preserved dinosaur turds contain snails that ate and lived inside the excrement. If you know what you're looking for, you can find tiny, tiny parasites in coprolites, too.

Those parasites have been found inside the petrified scat of carnivorous dinosaurs. In 2006, the paleontologists George Poinar and Arthur Boucot broke down a coprolite from an especially rich dinosaur site in Bernissart, Belgium. They scrubbed the specimen, ground it down into grains, suspended it in hydrochloric acid solution, spun the mixture in a centrifuge, placed it in hydrous hydrofluoric acid, centrifuged the products again, and so on, until the scientists had a concentrated dinosaur-poo residue that could be readily viewed under the microscope.

When Poinar and Boucot zoomed in, they found parasites. Tiny cysts indicated the presence of *Entamoeba*—a widespread genus of microorganisms that can be harmless or cause disease, depending on the species. And there were also eggs from both trematode and nematode worms. The prehistoric parasites were not identical to living species, but were similar enough that they could be identified. The little hitchhikers hadn't changed very much in 125 million years, and they indicated to paleontologists that dinosaurs were warm hosts to many familiar microorganisms.

Dinosaurs were assailed from the outside, too. While prehistoric lice are relatively few and far between, a combined approach using the known fossils and genetic data estimated that the major varieties of lice present today began to proliferate around 100 million years ago. Feathered dinosaurs had been around for at least 60 million years at that point, not to mention all the fuzzy mammals, and the boom among feather lice hints that the wingless insects had found plenty of homes. Dinosaur feathers were suitable homes for the parasites, just like the plumage of modern birds. With any luck, a sharp-eyed paleontologist will someday be able to pick one of these nits from the fossilized feathers of one of the beautifully preserved dinosaurs found in China's ash beds.

Dinosaur lice shared their world with ticks, mosquitoes, and other biting, burrowing insects. We don't know how many of these annoying invertebrates actually fed on dinosaurs (it's not like anyone's had the pleasure of catching one of the little biters in the act). But in the increasing array of possible parasites, there is one variety whose bloodsucking tools were so powerful that they would have been overkill on anything but a dinosaur. In early 2012, the Chinese Academy of Sciences paleontologist Diying Huang and his team announced that they had discovered huge, 165-million-year-old fleas. Granted, the fleas were only relatively gigantic— the largest was still just under an inch long—but to any potential victim, that's certainly big enough. Unlike their modern relatives, these fleas couldn't jump. They had heavy-duty mouthparts studded with saw-like projections, and this sturdy equipment led the researchers to propose that the tools went to work on dinosaurs.

The fleas were probably annoying ambush predators, they argued. The terrible fleas would wait for a dinosaur to amble by before scuttling out to latch onto their host, feed, and disappear back into the undergrowth.

Cataloging the list of dinosaur pests and parasites is a relatively new area of interest, though. For the most part, paleopathology has focused on injuries directly seen in dinosaur bones.

There's an entire book devoted to the various types of dinosaur pathologies, one that I like to occasionally flip through to remind myself of how dangerous dinosaur lives must have been. Organized by Tanke and the pathologist Bruce Rothschild, the volume carries the charming title *Dinosores: An Annotated Bibliography of Dinosaur Paleopathology and Related Topics—1838–2001*. Everything from biting insects to bone damage finds a place in the catalog. Highlights include early-twentieth-century speculations that dinosaurs got so large and weird because of glandular disorders, that a suite of bone fractures were attributable to rough dinosaur sex, and that they were poisoned by everything from arsenic and botulism to strychnine. The comprehensive list includes possible cases of spina bifida, osteomyelitis, necrosis, and gout.

Dinosaurs were even afflicted by cancer. Tumors both benign and malignant have turned up—not at such a rate that there's any indication that dinosaurs were suffering an increased cancer risk near their extinction, but we've indeed determined that this disease was present hundreds of millions of years ago. Tanke and Rothschild identified a benign tumor in the bone of a Jurassic dinosaur found in Utah, and in 1998 Rothschild and other scientists discovered metastatic cancer in another Jurassic dinosaur, this time from Colorado. Multiple cases have been found over the last ten years or so, most of them in hadrosaurs from the Late Cretaceous period.

Not all dinosaur diagnoses are on the mark. It's difficult enough to accurately identify disease today, much less in patients millions of years old (always get a second opinion from another paleontologist). Some of the earliest reports of dinosaur disease were influenced by the way scientists saw dinosaurs at the time. One of my favorite cases was described by the early-twentieth-century pathologist Roy L. Moodie. Even though paleontologists had noticed damaged dinosaur bones before, Moodie was one of the first researchers to fully catalog the wide array of pathologies found among fossil creatures in his influential book on the subject, simply titled *Paleopathology*. Among other fossils, Moodie highlighted two tail vertebrae from a sauropod similar to *Apatosaurus*. The size and shape of the bones showed that they came from near the end of the tail, but they didn't articulate with each other normally. A blob of inflamed tissue sat between two of the terminal tailbones. "The mass resembles closely the tumor-like masses seen on oak trees," Moodie wrote, and noted that such injuries could be seen in other dinosaur tails as well. Moodie invoked the sloth of the huge sauropods to explain how such an injury might have occurred. The tip of the dinosaur's tail "might be seized by one of the carnivorous dinosaurs and vigorously chewed for some time before the owner of the tail was able to turn his huge body and knock the offender away," he explained. Supposing that the sauropod's tail hadn't been significantly shortened, he reasoned, the bacteria that surely proliferated in the Jurassic swamps quickly infected the wound down to the bone.

Moodie's scenario was based on the commonly held idea at the time that huge sauropods were painfully stupid and were bound to live their lives in fetid swamps. He was wrong that sauropods were so sluggish and dim-witted that *Ceratosaurus* could simply bolt up to a sauropod tail for a quick and easy meal, but right that the tailbones of these dinosaurs were relatively delicate and prone to injury, and that in this and other ways their size and grandeur did not exempt them from what Shakespeare called "the thousand natural shocks / That flesh is heir to." Moodie observed, "The study of the lesions so far known among fossil animals indicates nothing new in the nature of pathological processes but simply extends our knowledge of disease to a vastly earlier period than had previously been known." While not all diseases present today existed in the Mesozoic, the signs of trauma and pathology among dinosaurs and other prehistoric creatures are familiar ones. The afflictions dinosaurs faced are still with us now.

The array of lesions, fractures, bite marks, fused bones, and other pathologies that Moodie, Rothschild, and other paleopathologists have identified show that dinosaurs were not impervious ultra-beasts. The fact that they were susceptible to injury and disease just like any other large vertebrate makes them all the more real. A pristine dinosaur skeleton, free of pathology, doesn't seem as authentic. A dinosaur with healed fractures or infected bones underscores the fact that the animal was once truly alive, and each injury goes back to some prehistoric event that we can at least trace the outlines of.

One dinosaur that I feel especially sorry for is Big Al, a teen *Allosaurus*. Like Sue and Jane, Big Al was an apex predator of his time, and suffered multiple injuries. Discovered in 1991 near the abundant Jurassic age Howe Quarry in Wyoming, Big Al was astonishingly complete, but his skeleton showed those telltale signs that he was badly beaten. The paleontologist Rebecca Hanna cataloged the dinosaur's injuries and found that he had *nineteen* abnormalities—from ribs and fingers damaged by trauma and infection to vertebrae damaged by more mysterious causes. None of the injuries directly caused the dinosaur's death, according to Hanna, but many of them hindered Big Al's ability to hunt. A fracture and resulting bone infection in Al's right hand were so bad that the dinosaur's second finger probably ached when he flexed it. And injuries on Al's left foot—still visible as an open abscess on his main weight-bearing toe bone, which would have suppurated, oozed pus, and *hurt*—hampered his ability to chase down prey. Big Al suffered a variety of painful injuries, but then again, the suite of pathologies also testifies to how resilient he was.

A roadside dinosaur stop just outside of Arches National Park records the steps of a dinosaur who knew Big Al's pain all too well. There is no sign or other hint that dinosaur ghosts are nearby; driving on State Road 191 past the towering red rock bluffs outside Moab, Utah toward I-70, the route to the tracksite is a dirt road that abruptly joins the main highway around mile marker 148.7. The first time I drove there, I wasn't entirely sure my little car was going to make it on the rough road. But I couldn't say no to dinosaur tracks.

I arrived at the little parking lot behind the low hills, and took the short hike up to the site, but I didn't see the footprints the interpretive sign told me I would. I walked back and forth, searching for them, and returned to the sign hoping that it contained some hint. Then I looked down—I was standing right on top of them. The easiest ones to spot were tracks left by a midsize sauropod. The gently curved tracks looked like an organized series of potholes. Impressive, but not the dinosaur traces I had come to see. The tracks I was looking for ran diagonally in the other direction—a series of large three-toed footprints most likely left by an *Allosaurus*. I was surprised to see that the tracks weren't spaced as they should have been. The dinosaur's steps alternated long and short. This carnivore was limping! The tracks record the movements of a hobbled *Allosaurus*, and as I found out, this injured dinosaur wasn't alone in its pain: sites from New Jersey to Australia record the movements of limping dinosaurs.

From broken bones to persistent parasites, injury, irritation, and disease were just a part of dinosaur life. Oddly enough, some paleontologists have turned to these clues to try to explain why the non-avian dinosaurs disappeared at the end of the Cretaceous. When Roy Moodie mapped out the pathologies in the fossil record, the injury curve peaked just before dinosaurs and other forms of Mesozoic life disappeared. Dinosaurs were having more and more accidents, Moodie implied, and "[i]t seems quite probable that many of the diseases which afflicted the dinosaurs and their associates became extinct with them."

Other paleontologists have tried to pinpoint specific causes—they've blamed cataracts, slipped discs, infectious epidemics, and even glandular disorders for the ultimate demise of the dinosaurs. But dinosaurs coped with disease and injury over the entire course of their history without any sign of slowing down or being winnowed into extinction. Something else must have extirpated them from the Earth. Why the non-avian dinosaurs went extinct is one of the greatest murder mysteries of all time.

Ten

Dinosaurs Undone

Of all the dinosaur mysteries, none is so confounding as why there are no descendants of *Tyrannosaurus* and *Triceratops* living alongside us today. And to understand that enigma, I knew I had to go to Montana—the resting place of some of North America's last dinosaurs.

When I finally got out to the dinosaur-rich exposures, though, I learned that cows are not the best company for fossil collecting. Even at a distance, spread out as splotches of black and brown over the Montana ranchland, the cattle, with their ceaseless lowing and grunting, were starting to break my concentration. I tried to tune them out and keep my focus on the ground (or at least keep myself from shouting expletives at the herd at the bottom of the hill). A momentary break in attention can make all the difference between spotting a piece of a dinosaur and stepping on it, unknowingly crushing it back into the ground.

I hadn't initially planned on doing much prospecting in Montana. I had driven out to the tiny town of Ekalaka from my last road trip stop in Bozeman for a different reason—an appointment with a specific *Tyrannosaurus rex*. A few months before my move out west, the Carthage College paleontologist Thomas Carr put out a call for field volunteers on Facebook, of all places.

Carr needed a few extra hands to help excavate a very young *Tyrannosaurus*—nicknamed "Little Clint"—that his team had discovered a few years before. I jumped at the chance and asked Carr for a spot. I'd be happy to dig up any dinosaur, but *Tyrannosaurus rex*? Count me in! Luckily, Carr responded that I'd be welcome to help.

But, as I was disappointed to learn, the young tyrant wasn't holding audiences. Soon after I rolled onto Ekalaka's quiet main drag—flanked by garbage cans decorated with sticker letters that spelled out "No Bird or Animal Parts, Thank You"—Carr and his field team delivered the bad news: an exceptionally wet winter had kept streams running high and wide, even into the late July heat. "There's a flooded stream between us and the Little Clint site that's about two city blocks wide," Carr lamented. It cut us off from the dinosaur, and I knew that you don't mess around with water in the field. I've extracted enough mired SUVs to know that even a brief cloudburst can turn smooth dirt roads into quagmires. Little Clint would have to wait for another season. At least the dinosaurian royalty wasn't going anywhere.

Fortunately, Scott Williams and his crew from the Burpee Museum of Natural History were also working in outcrops around Ekalaka. With Little Clint inaccessible, Carr decided to team up with Williams and look for other fossil sites. I'd have a chance to search for dinosaurs, after all.

Spread across the checkerboard of private ranches and Bureau of Land Management acreage around Ekalaka, exposures of the Hell Creek Formation present snapshots from the last days of the non-avian dinosaurs. The sediments, 65.5 million to about 66.8 million years old, record the reign of *Tyrannosaurus*, *Triceratops*, and *Edmontosaurus*, among others; but these dinosaurs and their close kin, as unstoppable as they might seem, all exited the evolutionary stage at the close of the Cretaceous. What's left of them rests silently beneath the hooves of grazing cattle and gradually tumbles out of hillsides.

On my first foray into the field on the Carr and Williams expedition, I asked the paleontologist Eric Morschhauser to point out the extent of the Hell Creek Formation so that I could get my bearings. I had read plenty about the Hell Creek, but had never worked there before, and it always takes a day or two for dinosaur hunters to develop a "search image" for the right layers of fossil-bearing rock. Morschhauser waved his hand out toward a wide valley covered in grass and dotted with hills and nubs of exposed rock. "All that out there," he said, "is Hell Creek." This was a land of untold opportunity. There were undoubtedly dinosaurs here. We just had to find them.

I got my chance a few days later. Williams led the combined ranks of the paleo teams to a bare, tan mound called simply "Scott's microsite." No one was going to find any perfectly preserved, articulated *Tyrannosaurus* skeletons here: microsites are spots where small, sturdy fossils are found in abundance—fish vertebrae, lizard jaws, and dinosaur teeth. They may not represent the kind of fossil hunting you see on television, but these aggregations of little fossils act as a census of who was around in a particular environment during a short period of time. So, with my bottle of water in hand and incessant mooing of livestock in my ears, I slowly started to scan the ground—looking for a little glint of enamel or color change that might give away a small bone. From a distance, the field crew must have looked like a crowd searching for someone's lost contact lens without stepping on it. It's really not that different.

Rain and wind had already done the hard work. Erosion had winnowed the fossils out of the soft sediment and left them all over the ground. Dinosaur teeth were the easiest to find. In an hour of collecting, I had picked up the isolated fang of a small *Tyrannosaurus*, a *Triceratops* tooth, and a few slightly curved teeth that belonged to the feathery, sickle-clawed deinonychosaurs. I found a flat spot to sit down and inspected the vestiges of dinosaurian lives long gone. Even though their avian cousins would survive—even thrive—in the millennia to come, every existing lineage of non-avian dinosaurs simply blinked out of existence. Not a single one was spared.

It's impossible to know who turned the lights out at the end of the Cretaceous world. We just don't have the geological resolution to piece together the final moments of that unfortunate creature. But, sitting on the fossil-rich hill, I imagine *Triceratops* as the final holdout. After all, the hulking, three-horned herbivore is more common than any other type of dinosaur in these deposits. You can hardly spend a day in the Hell Creek Formation without stumbling across some remnant of a *Triceratops* skull. So, technically, they had strength in numbers in the face of extinction, and could have been the species that defied extinction the longest. Turning over the single *Triceratops* tooth in my hand—just one component of the prodigious ceratopsian—I envision an old, solitary dinosaur standing in the Cretaceous twilight. One of her horns is broken, and her face is scarred with signs of a hard life in the land of tyrannosaurs. The terminal member of her species, she holds a lonely vigil as the horizon swallows the sun, and the Cretaceous closes. Or maybe, I think, that last *Triceratops* is a younger animal, confused that his herd has disappeared. He mournfully hoots into the dark night, waiting to hear a return call that never comes.

We're never going to know what happened. For all we know, the last non-avian dinosaur was one of the fluffy little sickle-claws, plucking strips of flesh from the carcasses of toppled behemoths. But no matter which species persisted the longest, that title is cold comfort given the ultimate extinction of all non-avian dinosaurs. You would think that at least some of these animals would survive. Dinosaurs were highly diverse and lived the world over. Their ultimate extermination is worse than a disaster—it is an abominable enigma reminding us that the universe is indifferent to life and evolutionary magnificence. If natural majesty made any difference in who survived and who disappeared, descendants of *Tyrannosaurus* and *Triceratops* should still hang on somewhere today. Extinction doesn't care a whit how astonishing prehistoric species seem to us.

Had the non-avian dinosaurs survived, though, I wouldn't exist to mourn their loss. A delicate fragment of fossil jaw made this clear. After my brief break, I slowly made my way over to where most of the other field crew members were searching. I got down on my hands and knees and gradually shuffled my way around the hill, checking the little divots and washes for any fossils that might have settled in them after the last storm.

I didn't even see the tiny skull fragment until I was right on top of it. The fragile fossil was beautiful—a chocolate-colored piece of an upper jaw, with two coal-black teeth still embedded inside. This was no dinosaur. The arrangement and anatomy of the teeth gave the bone away as part of a small mammal—a little beast with a differentiated tool kit of complex teeth covered in bumps, troughs, and cones.

The fossilized speck was the best thing I found that day. Dinosaur teeth are easy to find at microsites, but mammal fossils are very rare. As I turned the fossil over in my hand, I thought about my deep connection to the tiny critter. The jaw belonged to one of my Cretaceous family members, undoubtedly a fuzzy little thing with a twitchy nose surrounded by whiskers. This species lived in the dinosaur's world, yet some members of its mammalian kin didn't share the same fate as the great archosaurs. From my privileged perspective, I could look back and puzzle over why these meek little beasts eventually inherited the Earth.

Homo sapiens wasn't an inevitable evolutionary outcome. Our history has been molded by contingency and chance just as much as that of the dinosaurs. Even though our species is very young, at about 200,000 years old, our mammalian roots go back to diminutive insectivores that lived alongside some of the earliest dinosaurs. The delicate jaw I stumbled across among eastern

Montana's hills was one of those meek critters. Our ancestors and cousins were present for the entire dinosaurian reign. There was never any mammalian contest to dinosaurian dominance; no furry rebellion to overthrow the scaly and feathery rulers. The adaptability and ultimate success of the dinosaurs severely limited the evolutionary routes available to mammals. And although we might want to cheer for a few of our prehistoric relatives who ate dinosaurs—such as the badger-size carnivorous mammal *Repenomamus*, found with baby dinosaurs in its gut contents—mammals of the dinosaur era stayed small and inhabited the shadowy corners of the Mesozoic world.

The dinosaur decimation opened up ecological and evolutionary possibilities—opportunities exploited by survivors of a global disaster at the end of the Cretaceous, mammals among them. This dramatic, rapid turnover in life's history wasn't simply a matter of dinosaurs handing over the world to mammals. This was one of the worst mass extinctions of all time, an event that took with it the flying pterosaurs, reptilian sea monsters such as mosasaurs and plesiosaurs, the exquisitely coil-shelled cephalopods called ammonites, bizarre rudistid clams, and even some mammal lineages. What could have triggered such a tragedy?

The acceptance of mass extinctions as a reality of life on Earth, and one worth studying, took a circuitous path, thanks to preconceived notions about patterns of change on our planet. The early nineteenth century is a good place to start tracing this thread. During this time, the French anatomist Georges Cuvier divvied up the history of life into successive eras punctuated by dramatic catastrophes. Even though geologists had not yet determined a way to assign absolute ages to rock layers, naturalists of Cuvier's era perceived several major periods in the history of life on Earth. The first was dominated by fish and other sea life, the second was ruled by huge reptiles (such as the creatures that would eventually be called dinosaurs), and the third was ruled by mammals. There seemed to be clean breaks between successive eras, and Cuvier thought that sweeping calamities explained the sweeping changes to the sorts of life found in each era. Whatever species survived would then take over the vacated spaces and proliferate, and the next cataclysm would continue the cycle of devastation and renewal. (Unfortunately, Cuvier was never clear on what mechanism created the new forms of life that distinguished each age.)

Partially due to misunderstandings created by the religious and scientific views of his translators, though, Cuvier's work was associated with a kind of biblical literalism that was being washed away as geologists and paleontologists learned more about prehistory. A contrasting view, ushered in by geologist Charles Lyell and like-minded naturalists from the 1830s onward, considered the Earth to exist in a state of dynamic equilibrium in which gradual, stately change was the rule. There was no evidence for disasters of biblical proportion, and clinging to "catastrophist" views was chided as giving religion precedence over the evidence in the rock record. Paleontologists thought that life followed a similar tempo.

Charles Darwin's evolutionary theory—while not exactly welcomed by his peers when first announced—added to Lyell's "uniformitarian" viewpoint. Life gradually evolved at a steady pace, and as new species were born, they usually drove their parent species into extinction. Earth, and the living things upon it, were transformed in a logical, orderly manner. And since both Lyell's and Darwin's ideas were founded on the idea that changes in operation today entirely explained changes in the past—both geological and biological—there was no room for a global catastrophe on the scale of a mass extinction. No one had ever observed such a dramatic change, and to argue that life had been marked by such extreme pressures was special pleading.

Within this view of dignified, gradual change, the fact that dinosaurs disappeared wasn't all that surprising to paleontologists of the nineteenth and early twentieth centuries. As Michael Benton has emphasized, earlier generations of scientists thought that "the extinction of the dinosaurs was regarded as a minor hiccup in the progression of life." According to early-twentieth-century researchers such as Yale's Richard Swann Lull, dinosaurs "died a natural death" that was long overdue. "[T]he marvel is, not that [the dinosaurs] died," Lull wrote, "but that they survived so long." Nature had given dinosaurs a longer tenure than they deserved, and, according to paleontologists of Lull's generation, dinosaurs eventually succumbed to a built-in evolutionary self-destruct system that operated according to strange rules.

I was repeatedly told, from elementary school all the way through college, that scientists immediately accepted Darwin's evolutionary theory. But contrary to this textbook tale, naturalists didn't immediately welcome the notion of evolution by natural selection, especially paleontologists. They turned a critical eye on the idea: natural selection seemed too weak a force to modify organisms, and to those scientists who wanted to see the benevolent hand of God in nature, the mechanism seemed excessively violent and uncaring for the Creation. So, until the 1940s, when biologists from various backgrounds reaffirmed that natural selection was the main driver of evolutionary change, many paleontologists preferred alternative ideas: that there were internal forces that propelled organisms along golden roads of increasing perfection, or that there were vague biological drives that controlled the birth and death of species.

Lull subscribed to some of these ideas, and picked dinosaurs as a perfect example of evolution molded by mysterious internal forces. According to Lull, large size, a profusion of ornaments such as spikes and horns, and a general look of "degeneracy" were the telltale indicators of slipping evolutionary vigor, and dinosaurs showed all of these. Impressive as *Stegosaurus* and other armored dinosaurs appeared, Lull and other paleontologists presumed that the grotesque creatures were investing so much biological energy into growing horns and armor that they could hardly keep their other physiological and biological systems running. The same could be said of the biggest dinosaurs, like the high-shouldered, long-necked *Brachiosaurus*. These miserable dinosaurs were driven into extinction by evolutionary excesses, Lull and like-minded paleontologists believed, and the idea that dinosaurs grew too big and bizarre for their own good quickly spread through the public imagination. World War I protesters with the Anti-Preparedness Committee adopted Jingo the *Stegosaurus* as their mascot—a creature that was "All Armor Plate—No Brains," showing the dangers of stockpiling weaponry. And dinosaurs easily served anti-industrialists, too. The massive, undoubtedly dim-witted sauropods might just as well have been the titans of big business, while nimble small businesses—the economic equivalent of mammals—were poised to overrun the reptilian behemoths. Nobody wanted to "go the way of the dinosaur." This was the most embarrassing sort of extinction—an ultimate death from obsolescence and refusal to change.

George Wieland, a colleague of Lull's at Yale's Peabody Museum of Natural History, was one paleontologist who didn't follow this trend. Contrary to his early-twentieth-century colleagues, he didn't think dinosaurs were inherently doomed to fail; he argued, instead, that dinosaurs *ate* each other into extinction. Inspired by earlier hypotheses that small mammals frequently cracked open dinosaur eggs, Wieland proposed that giant lizards, snakes, and even tyrannosaurs and other theropods consumed so many sauropod eggs that the demand eventually outstripped the supply. Dinosaurs watched over their nests to drive off such thieves, Wieland speculated, but even the most loving of mothers would lose out to the egg-hungry hordes.

Wieland's papers were not the last word on dinosaur extinction. Not even close. Without a note informing them that "The butler did it," paleontologists were left to pick up a fossil cold case that had been lying dormant for 66 million years. The death of the non-avian dinosaurs was a scientific problem so nebulous, and so apparently irresolvable, that anyone who had a wild hair about the topic felt compelled to speak up. This long-running "dilettante" phase of idle speculation, as Michael Benton has called it, produced a slew of weird ideas that implicated just about every ailment or adverse environmental condition paleontologists and wannabe dinosaur experts could dream up. Global warming, global cooling, toxic plants, insufferable stupidity, cataracts, slipped discs, cosmic radiation, eggshells thinned by chemicals, and even reduced sex drive are just a few of the suspects out of more than a hundred ideas proposed during the twentieth century. As Benton has explained, "[T]he whole approach was apparently so easy and such fun that everyone felt that they had the opportunity, if not the duty, to solve the question of why the dinosaurs died out . . . It was as if, at the mere mention of 'dinosaur extinction,' scientists breathed a sigh of relief and felt freed from the straitjacket of normal scientific hypothesis-testing." There had to be some reason the dinosaurs disappeared, and the floor was open to just about anyone who had even the slightest inkling on the subject.

As far as I know, no one has yet implicated the CIA, the KGB, or Fidel Castro in the ultimate demise of the dinosaurs. That said, there has never been a shortage of truly outrageous hypotheses, including the sci-fi idea that aliens hunted dinosaurs out of existence. The idea is bunk, regardless of what the buffoons on *Ancient Aliens* might tell you, but is popular enough that the Utah State University Eastern Prehistoric Museum goes out of its way to list alien extermination along with disease and an ice age as unsupported hypotheses. ("There is no evidence of aliens or their garbage in the fossil record," the sign advises in deadpan lettering.)

But my favorite odd proposal was put forward by the entomologist Stanley Flanders in 1962. It is the David vs. Goliath (or, rather, Mothra vs. Godzilla) scenario of caterpillars versus dinosaurs. "The inherent weakness of the reptile was an extraordinary need for an abundance of plant material," Flanders wrote, and pointed out that Cretaceous caterpillars would have competed with the horned dinosaurs, hadrosaurs, sauropods, and other dinosaurian herbivores for food. Dinosaurs were big, sure, but caterpillars had strength in *numbers*. In his scenario, the insect larvae gobbled up the world's forests so quickly that a stunning profusion of butterflies eventually flittered over the carcasses of the starved dinosaurs. "Thus," Flanders concluded, "the giant reptiles which had survived during eons characterized by great changes in climate, continental uplifts, and different diets may have been exterminated by the lowly caterpillar." I guess we all have our favorite underdogs.

Of course, all of this idle speculation stems from the fact that the extinction of the non-avian dinosaurs and their unfortunate contemporaries was an event so massive in scale that it's difficult to comprehend what could have triggered such a tragedy. The conundrum has gotten only more frustrating with time. Dinosaurs were amazing creatures that, in one form or another, lived on this planet for 230 million years. They were not somehow due for extinction, or in decline—non-avian dinosaurs went out on top. In fact, as paleontologists have reconstructed dinosaurs with ever-greater clarity, and have realized that they were behaviorally complex and dynamic animals, the disappearance of *Triceratops* and company becomes ever more perplexing. What could have possibly wiped them out with such deadly discrimination? How could anything have destroyed them so mercilessly and completely? The list of possible killers has been narrowed down significantly, but it's one thing to find a possible murder weapon. It's another to figure out how and why it was used.

At the moment, the most prominent suspect is an asteroid, meteorite, or similar chunk of extraterrestrial rock that smashed into Earth's crust. There are a few competitors, mainly climate change, receding seaways, and intense volcanic activity. But none of these have the same public popularity. We know that all of these factors must have played some role. Intense scrutiny of the rock record has revealed that the world was drastically changing at the end of the Cretaceous. As the global climate cooled, the warm, shallow seas that straddled North America and other continents began to recede as ice built up at the poles. Simultaneously, a massive, sustained outpouring of lava at the Deccan Traps—which left wide flows of once-molten rock in what is now India—dumped greenhouse gases into the atmosphere, further altering climate and weather patterns.

All these things happened, and yet they've been overshadowed by what paleontologists call the "impact hypothesis." Gradual climate change never got its own big-budget natural-disaster flick, but the idea of extinction by extraterrestrial bolide got *two* mega-movies in the summer of 1998, not to mention the attendant Aerosmith single for *Armageddon*. There's something beautifully, destructively simple about an asteroid impact—the vision of dinosaurs being swept off the evolutionary stage by an unforeseeable stroke of misfortune. To borrow Neil Young's famous dichotomy, dinosaurs burned out rather than fading away.

How scientists detected the asteroid's impact has been told many times over. Still, it's worth recapitulating the discovery and the reaction to it, considering how radically the idea changed our perception of what happened to the dinosaurs. During the late 1970s, geologist Walter Alvarez was studying the latest Cretaceous rocks in Gubbio, Italy, when his attention was drawn to a half-inch layer of clay that he believed separated the reign of the dinosaurs and the beginning of the age of mammals. In other words, he speculated that the clay layer—known to paleontologists today as the K/Pg (Cretaceous-Paleogene) boundary—was deposited during the time of the end-Cretaceous mass extinction. Alvarez reasoned that he could measure the rapidity of the catastrophe if he could determine how long it took for that clay band to form.

Alvarez turned to his father, Nobel Prize–winning physicist Luis, to discuss the problem, and the elder Alvarez suggested a time-measurement technique that would ultimately spark a major controversy over the fate of the dinosaurs. Meteorites and other big-enough bits of cosmic flotsam and jetsam penetrate our atmosphere at a near-constant rate, he knew, and these alien rocks contain a relative abundance of elements rarely found in Earth's crust. One such element is a platinum-group metal called iridium. If this rare-earth element accumulated at a constant rate, the scientists reasoned, then they could use the amount of iridium in the clay layer to constrain the timeline for the end-Cretaceous extinction. As it turned out, the Gubbio clay layer was unusually rich in iridium. It wasn't an anomaly. When the Alvarez team investigated additional K/Pg samples from Denmark and New Zealand, they found that the spike was a consistent, real feature.

The abundance of iridium in the clay layers was *not* the result of an extended, slow period of deposition. Instead, in a 1980 *Science* paper, the Alvarez team concluded that the excess iridium came from an immense asteroid that, they hypothesized, struck the Earth right at the time of the Cretaceous mass extinction—the polar opposite of the slow, stately changing of the guard that paleontologists had suspected for decades was the real answer.

The Alvarez team wasn't the first to propose death from the skies, but they were the scientists who had the physical evidence to back it up. A few years before, the paleontologist Dale Russell and the physicist Wallace Tucker had proposed that a supernova triggered the end-Cretaceous extinction, but the Alvarez team had the answer locked in stone. The iridium band was the first undeniable clue that something strange had happened.

By the time my dinophilic self started pestering my parents to tape anything remotely dinosaur-related on TV in the mid-1980s, the problem of what caused the end-Cretaceous extinction seemed all but resolved. I knew this not because of Alvarez, but because Superman told me so. In 1985, Christopher Reeve (to me, the one and only Superman) hosted the Mesozoic extravaganza *Dinosaur!* The show was an inspiring mix of dinosauriana, from movie clips to notes on new discoveries and, best of all, superb stop-motion dinosaurs created by Phil Tippett. (While not entirely accurate, Tippett's creations still look far better than many of the stiff, cheaply rendered CGI dinosaurs that stampede across basic cable networks.) The show wasted little time setting the stage for the doom of the dinosaurs. Just under six minutes in, as an asteroid ominously rotates through the vacuum of space, Reeve asks, "What terror came out of the skies to end the dinosaurs?" Cut to a pair of courting *Edmontosaurus*—gentle and sensitive herbivores who guide their sole offspring through a hazardous Cretaceous world where a hungry tyrannosaur might be lurking just behind the next tree. These were not mindless abominations, but unique creatures with family values who were snuffed out by an awful happenstance.

The asteroid rocks the planet during the climax of the documentary. At first, not very much happens. A few trees fall and a *Tyrannosaurus* briefly loses its balance, and that's about all. Death slowly creeps up on the dinosaurs. A dust cloud blocks out the sun and kills off the vegetation, and a mother hadrosaur, alone in a scorched world, mourns over a nest of eggs that have all disintegrated. In time, she also disappears. And from beneath the bleached bones of a *Triceratops*, a hopelessly cross-eyed opossum waddles out into the sunlight to herald the first day of the Age of Mammals. The destruction of the non-avian dinosaurs wasn't triggered by a built-in factory flaw or geriatric decadence. Dinosaurs just had one really bad day from which they never recovered.

Almost every other documentary I saw during the 1980s and '90s replayed the same scenario. The dinosaurs were thriving until one fateful day when an asteroid struck the planet and wrapped the world in a cloak of dust, ash, and debris. Mammals, frogs, crocodiles, lizards, turtles, and birds were small enough to hide, but there was no hope for the dinosaurs. The lucky survived by virtue of adaptations they already possessed, and the unfortunate,

The extinction of the non-avian dinosaurs was a cosmic twist of bad luck.

(Cartoon by Mike Jacobsen)

lacking the variation necessary to survive, perished. The global devastation made beautiful, elegant sense. How could an asteroid approximately six miles across smack into the planet and *not* cause an environmental Armageddon? The end of the dinosaurs was not a gradual decline, winnowing down, or replacement by supposedly superior mammals. *Edmontosaurus* and its kind were cut down in their prime, the unsuspecting victims of cosmic chance. In our own era, the unfathomable devastation echoed Cold War fears of nuclear annihilation. Worries over "nuclear winter"— driven by the possibility of mutually assured atomic destruction— were fueled by the idea that dinosaurs had suffered such a fate millions of years ago.

•

Despite what I saw onscreen, most paleontologists in the mid-1980s bridled at the thought that the Cretaceous catastrophe was caused by a cosmic shock. This "impact hypothesis" looked like a deus ex machina that was far too simple to explain the death of so many different kinds of organisms all over the world. Climate change, the regression of Cretaceous seas, and ecological change following intense volcanic activity were the traditional extinction triggers. The fact that the Alvarez team came from disciplines outside paleontology didn't help the matter. Dinosaur experts spent countless hours picking over outcrops, studying the bones of the animals themselves, and trying to measure the pulse of life, and here come these snooty strangers to say that all that hard work had gone in the wrong direction.

Paleontologists resented the social and political fallout of the Alvarez debate. In 1985, a *New York Times* reporter relayed fears from paleontologists that papers critical of the impact hypothesis were being blocked by high-profile journals. Maintaining that changes in global climate, sea level, and habitat had killed the dinosaurs suddenly became a controversial point of view as the impact hypothesis grabbed the public's imagination. The idea was seductive, sensational, and didn't pay any heed to what paleontologists *thought* they were seeing in the fossil record. Paleontologist Robert Bakker was particularly incensed at the rapid acceptance of the hypothesis:

The arrogance of those people is simply unbelievable. They know next to nothing about how real animals evolve, live and become extinct. But despite their ignorance, the geochemists feel that all you have to do is crank up some fancy machine and you've revolutionized science. The real reasons for the dinosaur extinctions have to do with temperature and sea-level changes, the spread of diseases by migration and other complex events. But the catastrophe people don't seem to think such things matter. In effect, they're saying this: "We high-tech people have all the answers, and you paleontologists are just primitive rock hounds."

Bakker's feelings were echoed by many paleontologists—but, as in any debate, some pushed back and thought the Alvarez team had a point. One of the early converts to the impact hypothesis was the paleontologist David Raup, a specialist in large-scale patterns of evolution and decline over millions of years. His work was a central facet of a new paleobiology that promoted the study of fossils from what was perceived to be an intellectual backwater to one of the most essential components of evolutionary biology. While he admitted that he had an initial gut reaction against the impact hypothesis, Raup later changed his mind. In an essay about the debate (which continued for over a decade), Raup wondered why he himself had had such a negative reaction to the hypothesis. Why had paleontologists reacted with almost "universal derision" at the idea?

Tradition certainly played a role. "When I was training to be a paleontologist, in the 1950s," Raup wrote, "I was taught that most meteorite impacts on Earth were confined to what was known as the 'early bombardment,' the relatively short period of accumulation of debris left over from the formation of the Solar System." Comets, meteors, and asteroids were not supposed to be a regular feature of Earth history, and paleontologists largely ignored the growing body of evidence that such bodies, sometimes of considerable size, had continued to fall to the planet. The end-Cretaceous impact wasn't an unusual event that required special pleading for its possibility. The idea fit in with the long history of impacts on the Earth.

And, not long after the Alvarez team made their contentious proposal, the actual impact crater was finally recognized. The 111-mile-wide depression in the Earth's crust—situated in the vicinity of the Yucatan Peninsula—had been discovered years before by oil geologists, but no one understood the pit's significance until a decade after the asteroid controversy started. In 1991, the geologist Alan Hildebrand and coauthors tied the crater to the iridium evidence the Alvarez team found, and the geologic wound was dubbed the Chicxulub Crater. Based on its location, the asteroid strike was close enough to the home range of *Tyrannosaurus*, *Triceratops*, and other famous end-Cretaceous dinosaurs to wipe them out instantaneously. (As my Paleontology 101 professor, William Gallagher, once told my class, the impact would have created a devastating wave of fire, steam, and debris that would have "flash fried" the dinosaurs of North America.) There could no longer be any dispute about whether the impact happened. Suddenly, the pattern of life on Earth seemed very different.

Until the impact debate, most paleontologists had long upheld the Darwinian view of extinction as an outcome of the struggle for survival. Novel, better-adapted forms outcompeted and replaced their predecessors, and so the flow of life—while brutal—remained orderly. This traditional view was fading away thanks to a growing body of knowledge about the fossil record and the creatures themselves. There were no obvious signs that mass extinction victims were objectively inferior or that survivors were absolutely superior. Any number of dinosaurian liabilities could be—and had been—concocted to explain away their extinction, but, as Raup noted, just as many "biological failings" could have been found among mammals if the extinction had turned out differently. (And, according to a recent study by Mark Norell and colleagues, dinosaurs were still thriving at the very end of the Cretaceous. There was no global dinosaur decline that might indicate that they were already on their way out.)

This was one of the most unsettling aspects of the impact hypothesis. Death in a mass extinction was partially attributable to luck; a species could not foresee or prepare for the test of survival. The natural history of some species would make them more vulnerable to extinction while others would be more resistant. Yet identifying a possible extinction trigger isn't the same as showing how that event actually wiped out a species. We know that an enormous asteroid struck Earth around 66 million years ago, but what happened between that instant and the death of the last non-avian dinosaurs?

Almost everything we know about the evolutionary changing of the guard comes from a relatively small area of Montana and adjacent states. We're still piecing together what happened. The mass extinction was undoubtedly a global phenomenon—there is no solid evidence that dinosaurs survived into the Paleocene at any site in the world—but most of what we know about life just before the disaster comes from this narrow swath of western North America. The reason why doomsday scenarios in dinosaur documentaries invariably focus on *Tyrannosaurus, Triceratops, Troodon,* and *Edmontosaurus* as the stars is not just because these dinosaurs are among the most famous, but because this assemblage of dinosaurs is the only terminal Cretaceous community that has been extensively studied so far. There are other end-Cretaceous sites in Europe and Mongolia, and there may be more around the world, but so far we've only just begun to reconstruct what the entire world was like just before the asteroid hit.

Without clear and comprehensive Before and After shots of the end-Cretaceous transition, it's extremely difficult to detect the secret of the non-avian dinosaurs' demise. We know when it happened, and we now know it happened fast. Yes, an asteroid struck the planet at a time already assailed by massive volcanic outpouring, climate change, and receding sea level, but how did any or all of these different factors translate into pressures that wiped out my favorite creatures? It's not enough to simply correlate a mass extinction with a particular event. We need to gauge how a specific cause or causes triggered an ecological cascade ending in the survival or demise of a specific group of animals.

The asteroid impact is such an unfamiliar and sensational event that it will undoubtedly dominate the discussion for years.

Based on our current evidence, it's the simplest explanation available for why dinosaurs exited what the paleontologist W. D. Matthew once called life's "splendid drama." All the same, the debate over what wiped out the dinosaurs isn't over.

Nothing summarized the current state of affairs better than a recent academic kerfuffle sparked by a 2010 *Science* paper that reaffirmed the impact hypothesis as the chief cause of the end-Cretaceous mass extinction. More than forty geologists and paleontologists collaborated on the position paper, which concluded that the vast weight of evidence pinned the responsibility for the extinction on an asteroid impact. Not all paleontologists agreed. Soon after, *Science* published dissenting opinions from several other groups of scientists. Dinosaur specialists and other vertebrate paleontologists, in particular, said that the asteroid impact probably played a major role in the extinction, but that other causes—such as sea level change and volcanic activity—could not be discounted.

In light of this exchange, I called a few paleontologists to get a read on the current state of the debate. The University of Rhode Island paleontologist David Fastovsky, who has been a vocal advocate of the impact hypothesis for years, told me that the global pattern—on land and in the seas—points to a single, catastrophic cause. An asteroid impact best fits the terrible trend. An appeal to multiple causes, Fastovsky said, "philosophically bothers me" because "they only explain a piece of the data" rather than the big picture. Thus far, the idea that an asteroid impact was the chief culprit has survived the myriad tests and criticisms thrown its way.

Mark Goodwin took a different view when I asked him about the extinction. The impact hypothesis is "a sort of hook to hang one's hat on for the demise of the dinosaurs," and it "certainly caused some stress in the environment," Goodwin said, but we don't know enough yet to determine whether it entirely explains the massive destruction at the end of the Cretaceous. Goodwin likened the state of affairs to the current scientific investigation of climate change. "We all agree that the world is warming up" and that humans are behind it, Goodwin said, "but we disagree on the tempo and mode." The points of contention are not major theoretical differences, but "fine points" that require additional data and refined techniques that we don't yet have access to.

I called the Mesozoic mammal expert Anne Weil, too. After all, some mammal lineages were cut back or died out at the end of the Cretaceous. Perhaps, by studying the record of our tiny relatives, we can better understand what happened to the dinosaurs and other animals present at the end of the Cretaceous. "This was the biggest event in mammalian evolution," Weil said. "It was amazing." About half the mammals present at the end of the Cretaceous were lost, and the ones we find on the other side of the extinction—at the dawn of the next geological era—formed the basis for the beasts we see around us today. Determining what happened in that small slice between the end of the Cretaceous and the earliest part of the Paleocene is the trick. We simply don't have the high-definition data to tell whether the extinction took place over the course of weeks or thousands of years, Weil said. Since we presently lack the refinement to track how ecosystems changed in the days, months, and years after the asteroid strike, we can't confidently say that the impact was the one and only extinction trigger. "I don't have any trouble with there being an impact," Weil said, "but show me how it works."

More than thirty years after the impact hypothesis reinvigorated the debate over what happened to the non-avian dinosaurs and so many other life forms that shared their world, we're really only just starting to understand what happened at the end of the Cretaceous. Speaking for myself, I believe that geologists and paleontologists have pinpointed the major players in the extinction, but we have a long way to go before we can tease apart how these factors contributed to one of the most devastating episodes in the history of life. We still lack the delicate resolution needed to comprehend why the Cretaceous went out with a bang. With so much left to discover, I'm in no rush to declare the case closed.

All we can say for certain about that great extinction is that the swift desolation forever changed the history of life on Earth. As Weil pointed out, the event gave mammals their big evolutionary chance, and the loss of almost all the dinosaur lineages irrevocably changed our world. Dinosaurs didn't just happen to live on this planet. They were interacting parts of connected ecosystems, and their long tenure on Earth helped set the stage for life as we know it today.

The paleontologist Jack Horner once said that he wasn't interested in how dinosaurs died. He was interested in how they lived. He said this at the height of the extinction controversy, in the 1980s. I can't blame him for wanting to just stay out of the whole mess. But dinosaurs' lives and their ultimate demise are inextricably tied together. If we can better understand the biology of dinosaurs, we can start to figure out why avian dinosaurs survived while the most monstrous forms were lost forever. Mulling over ideas about dinosaur extinction will lead to discoveries, yielding information on aspects of dinosaur biology we hadn't expected. The more we learn about dinosaurs, the more complex they become. Even as we find answers, we're met with new questions. If dinosaurs survived millions of years of climate change, continental drift, and even other mass extinctions, then why did most dinosaurs perish 66 million years ago? Dinosaurs certainly had the flexibility to survive. Some of them did, in feathery avian garb. The riddle is why the non-avian sort—the ones that stalk our dreams and nightmares—did not.

Epilogue:
My Beloved "Brontosaurus"

When I was little, all I wanted was a pet dinosaur. I kept hoping that someone would find one in a distant jungle, or, taking a cue from Michael Crichton's fictional genetic engineers, that scientists would hurry up and find a way to bring dinosaurs back to life. I had no idea that Sweetie—my pet budgie—was a dinosaur, and, even if I had known, she wasn't the same as a sauropod I could ride to school every day. It was unfair that dinosaurs were lost to history. I wanted the dinosaurs to come back. So now, over twenty years after I first met my *"Brontosaurus,"* it feels a little strange to say that I'm glad that the most wonderful and awesome of dinosaurs are extinct.

Sometimes I still imagine parting the curtains one morning to find an *Apatosaurus* plucking leaves from the tree on my front lawn. I'll admit that, and I wouldn't look a gift dinosaur in the mouth. But my attitude has changed since I was a kid. I know that dinosaurs do more for us dead than alive. Just like Obi Wan Kenobi in *Star Wars*, when he warned Darth Vader that he would be more powerful in death than in life, the lessons and wisdom we gain from dinosaurs could have come only from their tragic demise.

If *Apatosaurus* and other dinosaurs had been given a stay of execution, they wouldn't seem so special to us. Even though we know birds are dinosaurs, we don't cherish them the way we do their Mesozoic relatives. They're too familiar. The same is true of the bizarre and varied fossil mammals that thrived in the wake of the end-Cretaceous extinction. They're just as spectacular as dinosaurs, but they're too close to animals we see around us today. A large part of why we keep going back to dinosaurs, I think, is because they're so alien and exotic. There hasn't been anything quite like them in 66 *million* years. The deep gulf created by extinction, separating us from them, makes dinosaurs all the more fantastic. Just think of all of our favorite dinosaurs—*Apatosaurus, Tyrannosaurus, Triceratops, Stegosaurus.* There is nothing in our era that comes close to their magnificence.

Time matters. Let's say a descendant of *Apatosaurus* had held on, bucked the trend, and persisted to the present day. We could observe, study, and carefully dissect that dinosaur to solve any number of mysteries about its prehistoric counterparts. But we don't have that straggling sauropod thread to draw answers from. At least the truth about dinosaurs is kept safe in their remains. If we're going to understand how dinosaurs like *Apatosaurus* lived, we have no choice but to carefully and gradually piece together what we can from their petrified vestiges. The most amazing of dinosaurs come to us as stone outlines of what they once were, and the struggle of science is to interpret their bones.

For too long, we saw dinosaurs as prehistoric sideshow freaks. Dinosaurs used to be symbols of reptilian excess that deserved extinction. When scientists falsely believed that evolution followed the path of Progress—with our species sitting at the glorious apex and end of evolution—they saw dinosaurs as a weird, if charismatic, interlude in the story of life. Now we know better. Dinosaurs are not just icons of extinction and prehistoric lineages snuffed out. They are the grandest of Darwin's "endless forms most beautiful and most wonderful," prominent players on the evolutionary stage that instantly cause us to question our own place in nature and in the history of life on this planet.

Cultures all over the world discovered prehistoric bones before paleontologists began to catalog all the dinosaurs we know and love. People recognized the relics for what they were—remains of creatures that once lived—and concocted legends about the monsters, heroes, and gods. Maybe the particulars of their conclusions don't fit neatly into what we've discovered since science distilled the truth from those myths, but the important fact is that dinosaurs demanded an explanation. Long before we had a word for them, dinosaurs made us question what happened in prehistory, and what those ancient skeletons might be able to tell us about how the modern world came to be.

The shadows of ancient lore have only recently given way to the real animals, which are even stranger than anything we could have imagined. This is the other great secret of the dinosaurs. Even though we might complain about the loss of *"Brontosaurus,"* or feel that a feathery *Velociraptor* is not a *Velociraptor* at all, fixating on these sore spots obscures just how much we're learning about dinosaurs. If you really look at the evidence, there is no question that the dinosaurs we're bringing to life today are far more beautiful and complex than their earlier incarnations. What is more amazing? A *"Brontosaurus"* forever bound to a slow-motion life wallowing in a scummy Jurassic pond, or a herd of taut, brightly colored *Apatosaurus* stomping across fern-covered floodplains, their sinuous tails twisting in the air as they walk in formation? Dinosaurs are better than ever. *"Brontosaurus"* is better left in the past—a warm reminder of dinosaurs as I once knew them, and of how far our understanding has come.

A friend once asked if I was angry at scientists for taking away my childhood dinosaurs. It's complicated, I replied. I don't feel any ill will toward the loss of the dumb olive-green sluggards that populated library books and flickered across the television set when I was young. The dinosaurs I first met have been torn limb from limb by visions of far more colorful, active, and, honestly, interesting dinosaurs. Maybe I was just young enough that I soaked up the contemporary dinosaur revisions without a second thought. I'm not entirely callous. Every now and then, I pull a woefully outdated dinosaur book off the shelf and spend a few minutes envisioning the goofy cold-blooded reptiles that first sparked my love for the past. I remember them fondly, but I don't need *"Brontosaurus"* and other old-school dinosaurs to come back. I'm content to leave her in my childhood past. I love her from afar, and keep the sauropod safe in my memory.

But there is another reason not to forget the dinosaurs of my youth: even outdated dinosaurs can share secrets.

Of all the dinosaurs I've ever seen, none encapsulates my feelings better than the *Apatosaurus* at Yale's Peabody Museum of Natural History. I had a chance to visit the old dame in the fall of 2010, when I skipped out on a conference to spend some time with the institution's dinosaurs before the evening reception crowded the exhibit hall with colleagues sipping cheap wine from plastic cups. Standing in front of the gargantuan, long-necked skeleton, I felt like a kid again. The gloomy hall reminded me of my early visit to the AMNH and other dark museums populated by shadowy dinosaur frames. But I knew much more than I did when I was a kid. For one thing, I knew that the bones used to make the mount were the ones O. C. Marsh's field workers collected from Como Bluff, Wyoming, so long ago. I knew the dinosaur's proper name full well, but I couldn't deny the title Marsh himself had once bestowed on these bones. This was the original *"Brontosaurus."*

After Marsh's death, when the museum set about displaying the dinosaurs for the public, they erected the sauropod according to the science of the time—as a dull, tail-dragging behemoth. An expansive mural that runs along the exhibit's wall underscores this vision. A gorgeous fresco secco painted by artist Rudolph Zallinger in the late 1940s, the *Age of Reptiles* restores the monumental dinosaur in a Jurassic swamp. I had seen facsimile posters and prints before, but those reproductions did not do justice to the beauty of the original. *"Brontosaurus"* herself is stunning. The sun glints off her iridescent scales as she strains soft plants from the ancient pool. The mural has remained a monument to the vintage image of *"Brontosaurus"* all these years, even after paleontologists changed the skeleton below to fit the dinosaur's new identity. Held high off the ground by the dinosaur's intricate neck is the proper *Apatosaurus* skull—placed on the skeleton during the great dinosaur revival of the 1970s.

This skeleton, just by itself, records more than half the history of paleontology. Our ever-shifting image of dinosaurs—from the days of Marsh to the Dinosaur Renaissance—formed a historical mosaic in this one dinosaur. The Yale *Apatosaurus* isn't an image of prehistoric life perfectly assembled the first time and left to stand. The Police were dead wrong about this dinosaur—there are lessons in the past of the mighty *"Brontosaurus."*

The dinosaur may keep changing. Beyond the tweaks and updates to what we know of the dinosaur's biology, rumor has it that *"Brontosaurus"* may be revived someday. According to fossil gossip, two special skulls show that *Apatosaurus excelsus* was markedly distinct from the other two *Apatosaurus* species. If this is true, and is confirmed by future studies, paleontologists could make the case that *Brontosaurus excelsus* should be revived.

The name controversy is just a footnote to the vibrant story of how paleontology has investigated the lives of dinosaurs in ever-greater detail. Since the time we first met the dinosaur, our image of *Apatosaurus* has evolved from a reptilian caricature to a detailed portrait of a real, spectacular animal unlike anything since the end of the Cretaceous. And that's the aspect of the old dinosaurs I cherish most: they record how our perception of the dinosaurian essence has evolved. I am thrilled and fascinated by the new dinosaurs, and often frustrated by out-of-date skeletons and museum displays, but at the same time I feel grateful for them. Contrasted with what we know now, and placed into the context of history, the old dinosaurs can show us just how much our understanding has transformed. The *Apatosaurus* isn't just a beautiful specimen. It's a potent embodiment of how science works—the interaction of fact, theory, and imagination that lets us approach animals we will never see alive.

What we know today will be tested and questioned by what we find tomorrow. And I am thrilled to live during a time when paleontologists are interrogating the fossil record with ever more precision. Today's images will slowly fade as the new dinosaurs come into view, and our perception of what makes a dinosaur a dinosaur will subtly shift from one generation to the next. When I was a kid, for example, the idea that any dinosaurs had feathers was a speculative and heretical notion. Today we know that most lineages of dinosaurs may have been covered in bristles and fuzz. As scientists and artists bring this under-

A *Deinonychus* leaps from the shadows at Yale's Peabody Museum of Natural History. The juxtaposition of this "hot-blooded" dinosaur with the outdated mounts elsewhere in the hall illustrates just how much our understanding of dinosaurs has changed. (Photograph by the author)

standing to life, the next generation will grow up with fluffy dinosaurs.

If I've learned anything during my time in the company of dinosaurs, it's that we desperately need to keep them in mind as we watch our world change around us. Shortly before I finished the research for this book, I took one last trip out into the field. Thanks to an unseasonable warm snap, I had the chance to go out to Grand Staircase–Escalante National Monument in mid-March—a time when snow and chilly temperatures often hamper attempts to go fossil hunting. Two University of Utah graduate students and I were meeting up with the Denver Museum of Nature and Science field crew to search for Late Cretaceous dinosaurs.

This is an exceptionally rich fossil fantasyland. In other places, like the Triassic outcrops of Dinosaur National Monument, dinosaurs are so rare that any scrap of bone or tooth fragment is a significant find. But here, dinosaurs and their neighbors are so common that fossil hunters have the luxury of picking and choosing what to collect. The dry hills are littered with bone, which can actually prove quite frustrating. I always feel a little rush when I spy fossil scraps peeking out of a hill, but here, many of those dinosaur signs are bones that recently decayed into dozens of petrified shards from prolonged exposure to sun, wind, and rain.

On the second morning of our survey, I gasped for breath trying to keep up with Ian Miller, the Denver Museum's paleobotanist, as we hiked out to the drop-in point for the basin we had come to explore. I wasn't dragging as badly as a few members of the team who had overindulged in cheap booze around the campfire the night before (you know it's quality whiskey when it comes in a big plastic bottle), but all the water I carried in my pack weighed me down enough that walking while talking dinosaurs was a serious test of endurance.

As we passed twisted junipers along the dirt and cobblestone road, I asked Ian what the local environment looked like when the many-horned *Kosmoceratops* and the short-snouted tyrannosaur *Teratophoneus* roamed there. Ian explained that it was very different from what we looked out at that day. Seventy-five million years ago, during the time preserved in the Kaiparowits Formation, this swath of southern Utah was a wet, warm coastal dinosaur paradise. Instead of cracked tamarisk and sage, the land was carpeted in lush vegetation that more closely resembled a modern rain forest—tall stands of vine-draped trees stood between patches of swamp. What is now southern Utah looked like coastal Louisiana, just with dinosaurs and far bigger alligators.

North America's shallow sea was part of what allowed Utah's southern dinosaurs to evolve in splendid isolation, and it drained off the continent just before the curtain fell on the Age of Dinosaurs. After the cataclysm, rain forests filled with archaic primates and strange varieties of mammals covered the land, and, after another 50 million years of climate change, continental drift, erosion, and uplift, Utah's badlands were formed. Even now, the world's metamorphosis continues. It's at a rate that's practically impossible to detect with our own eyes, but it's happening. The scale of a human life—measured by the speed of Internet updates or the crawl of a working day—is ill suited to fit the dynamic nature of our planet and the fantastic organisms that continue to evolve here.

We're influencing those changes. As narrow-minded naysayers continue to deny the frightening weight of evidence that we are changing the global climate and hastening the extinction of untold species, the harmful effects of our industry and technology continue to build. No amount of stubborn denialism is going to alter the fact that we are transforming the world so quickly that even if we stopped pumping greenhouse gases and dumping pollutants today, the hallmark of our destructive nature would still be visible on the world for centuries to come. While we debate, the world changes.

I reflected on this while we walked from camp to the point

where the fossil-rich Kaiporowits opened before us. The world is changing so drastically, so quickly, and here I am sifting through the past. A persistent and pernicious doubt wormed its way to the front of my mind: why should we care at all about dinosaurs?

As I walked down the narrow path toward the basin, I kept turning the question over in my mind. Dinosaurs really do matter, I argued back to myself. The fossil record has taught us powerful lessons that we ignore at our peril. Pick any dinosaur you like, and that ancient creature is undeniable proof that our planet has a history so deep that we can barely comprehend it, that life has changed dramatically over time, and that extinction is the ultimate fate of all species. Nothing so majestically encapsulates these simple, powerful truths of nature quite like a dinosaur.

By investigating what dinosaurs really were, we put our own history in context—especially since our history is interwoven with theirs. Our ancestors and relatives lived alongside dinosaurs from the very beginning, and the evolutionary history of our mammalian kin was irrevocably influenced by dinosaur dominance. This is the strange duality of the dinosaurs. Mammals thrived once non-avian dinosaurs left the scene, one of many evolutionary events that made our unexpected origin possible, but we probably wouldn't be here if the dinosaurs had never existed. During all those millions of years that dinosaurs reigned, our forebears proliferated and evolved in the shadows, and these timid beasts set the foundation for the mammals that would eventually evolve—including our own primate lineage. Our history is not separate from dinosaurs. We share a deep connection with them.

I was left with plenty of time to think about dinosaurs as I spent the rest of the day scouring the remote basin. I have loved dinosaurs for as long as I can remember. They have always been there, lurking in the background if not more prominently striding through my mind as they do now. I started off imbibing the imagery, using skeletons and illustrations as launching points for my own dreams. But now, as I scuffed over hills covered in crumbling bones and tried to pick out the tantalizing glint of dinosaurs just barely poking out of the rock, I got the rare opportunity to experience something beyond what I imagined. I was searching for the rare, beautiful remnants of a lost world that could tell us just a little bit more about the rise and fall of the these astonishing creatures. If I could find a dinosaur, and rescue it from the rock, who knows what secrets it might eventually unlock?

To many, dinosaurs may seem to be childhood kitsch and fantastic monsters. But without them, we would not be what we are. Dinosaurs are instantly recognizable icons of evolution and extinction—triumphant and ultimately tragic creatures that beautifully illustrate the duality of life's contingent thread. They are guideposts to the past and harbingers of what the future might bring. We need dinosaurs.

Notes

1. Dragons of the Prime

12 **This open, airy vibe:** Lowell Dingus, *Next of Kin: Great Fossils at the American Museum of Natural History* (New York: Rizzoli, 1996).

14 **The story started:** Keith M. Parsons, *Drawing Out Leviathan: Dinosaurs and the Science Wars* (Bloomington: Indiana University Press, 2001), 1–21.

16 **As the AMNH paleontologist:** William Diller Matthew, "The Mounted Skeleton of Brontosaurus," *American Museum Journal* 5, no. 2 (1905): 63–70.

17 **The protected excavation:** Brian Switek, "America's Monumental Dinosaur Site," Smithsonian.com, May 31, 2012, www.smithsonianmag .com/science-nature/Americas-Monumental-Dinosaur-Site.html.

19 **In September of 1909:** Parsons, *Drawing Out Leviathan*, 1–21.

21 **In 1975, a physicist:** John S. McIntosh and David S. Berman, "Description of the Palate and Lower Jaw of the Sauropod Dinosaur *Diplodocus* (Reptilia: Saurischia) with Remarks on the Nature of the Skull of *Apatosaurus*," *Journal of Paleontology* 49, no. 1 (1975): 187–99.

21 **Yale's Peabody Museum exchanged skulls:** "Yale Brontosaurus Gets Head On Right at Last," *New York Times*, October 26, 1981, www .nytimes.com/1981/10/26/nyregion/yale-brontosaurus-gets-head-on -right-at-last.html.

22 **a previously unknown sauropod, *Brontomerus*:** Michael P. Taylor, Mathew J. Wedel, and Richard L. Cifelli, "A New Sauropod Dinosaur from the Lower Cretaceous Cedar Mountain Formation, Utah, USA," *Acta Palaeontologica Polonica* 56, no. 1 (2011): 75–98, doi: dx.doi.org/10.4202 /app.2010.0073.

22 **Just look at Google's Ngram Viewer:** Google books Ngram Viewer, accessed July 13, 2012, books.google.com/ngrams/graph?content=Bronto

saurus%2C+Apatosaurus&year_start=1800&year_end=2012&corpus=0&
smoothing=3.

23   **As the astronomer:** Mike Brown, *How I Killed Pluto and Why It Had It
Coming* (New York: Spiegel & Grau, 2010), xii.

24   **The dyspeptic Victorian anatomist:** H. G. Seeley, "On the Classifi-
cation of the Fossil Animals Commonly Named Dinosauria," *Proceedings of
the Royal Society of London* 43 (1887–1888): 165–71.

26   **As the journalist John Noble Wilford:** John Noble Wilford, *The
Riddle of the Dinosaur* (New York: Alfred A. Knopf, 1985), 168.

26   **NASA even shot:** Brian Switek, "Dinosaurs in Space!," *Dinosaur Tracking*, Blogs, Smithsonian.com, December 12, 2011, blogs.smithsonianmag
.com/dinosaur/2011/12/dinosaurs-in-space/.

27   **From the Greeks to Native Americans:** See Adrienne Mayor, *Fossil
Legends of the First Americans* (Princeton: Princeton University Press, 2005).

28   **As George Gaylord Simpson:** See George Gaylord Simpson, *Attending
Marvels: A Patagonian Journal* (New York: Macmillan, 1934; Time-Life
Books, 1965, 1982), 82.

29   **"My dinosaur is bigger than yours" contest:** See Paul D. Brinkman, *The Second Jurassic Dinosaur Rush: Museums and Paleontology in America at
the Turn of the Twentieth Century* (Chicago: University of Chicago Press, 2010).

2. *The Secret of Dinosaur Success*

36   *Revueltosaurus* **wasn't a dinosaur at all:** William G. Parker, Randall B. Irmis, and Sterling J. Nesbitt, "Review of the Late Triassic Dinosaur
Record from Petrified Forest National Park, Arizona," in *A Century of Research
at Petrified Forest National Park: Geology and Paleontology*, ed. William G. Parker,
S. R. Ash, and Randall B. Irmis, Museum of Northern Arizona Bulletin no.
62 (Flagstaff, Arizona: Museum of Northern Arizona, 2006), 160.

43   **The paleontologist Alan Charig:** Alan J. Charig, "The Evolution of
the Archosaur Pelvis and Hind-Limb: An Explanation in Functional
Terms," in *Studies in Vertebrate Evolution: Essays Presented to F. R. Parrington*,
eds. Kenneth A. Joysey and Thomas S. Kemp (New York: Winchester
Press, 1972), 121.

44   **an article entitled "The Superiority of Dinosaurs":** Robert T.
Bakker, "The Superiority of Dinosaurs," *Discovery* 3, no. 2 (1968): 11–22.

44   **mammals were runners-up:** Robert T. Bakker, "Dinosaur Physiology and the Origin of Mammals," *Evolution* 25, no. 4 (1971): 636–58.

46   **the work of the early archosaur expert:** Sterling J. Nesbitt, "The
Early Evolution of Archosaurs: Relationships and the Origin of Major
Clades," *Bulletin of the American Museum of Natural History* 352 (2011), dx.doi
.org/10.1206/352.1.

48   **The *Coelophysis* bonebed:** See Edwin H. Colbert, *The Little Dinosaurs
of Ghost Ranch* (New York: Columbia University Press, 1995).

50 **Sterling and his advisor Mark Norell named the animal *Effigia okeeffeae*:** Sterling J. Nesbitt and Mark A. Norell, "Extreme Convergence in the Body Plans of an Early Suchian (Archosauria) and Ornithomimid Dinosaurs (Theropoda)," *Proceedings of the Royal Society B* 273, no. 1590 (2006): 1045–48, doi:10.1098/rspb.2005.3426.

51 **their close relative *Poposaurus*:** Jacques A. Gauthier et al., "The Bipedal Stem Crocodilian *Poposaurus gracilis*: Inferring Function in Fossils and Innovation in Archosaur Locomotion," *Bulletin of the Peabody Museum of Natural History* 52, no. 1 (2011): 107–26, dx.doi.org/10.3374/014 .052.0102.

53 **Some of the earliest known archosaurs:** Richard J. Butler et al., "The Sail-Backed Reptile *Ctenosauriscus* from the Latest Early Triassic of Germany and the Timing and Biogeography of the Early Archosaur Radiation," *PLoS ONE* 6, no. 10 (2011), doi:10.1371/journal.pone .0025693.

53 **dinosaur forerunners were walking:** Stephen L. Brusatte, Grzegorz Niedźwiedzki, and Richard J. Butler, "Footprints Pull Origin and Diversification of Dinosaur Stem Lineage Deep into Early Triassic," *Proceedings of the Royal Society B* 278, no. 1708 (2010): 1107–13, doi:10.1098 /rspb.2010.1746.

54 **Recent discoveries such as *Asilisaurus*:** Sterling J. Nesbitt et al., "Ecologically Distinct Dinosaurian Sister Group Shows Early Diversification of Ornithodira," *Nature* 464, no. 7285 (2010): 95–98 doi:10.1038/nature 08718.

54 ***Eoraptor* may have been more of a plant-focused omnivore:** Ricardo N. Martinez et al., "A Basal Dinosaur from the Dawn of the Dinosaur Era in Southwestern Pangaea," *Science* 331, no. 6014 (2011): 206–10, doi:10.1126/science.1198467.

54 **the razor-jawed *Saurosuchus*:** Oscar Alcober, "Redescription of the Skull of *Saurosuchus galilei* (Archosauria: Rauisuchidae)," *Journal of Vertebrate Paleontology* 20, no. 2 (2000): 302–16, dx.doi.org/10.1671/0272-4634 (2000)020[0302:ROTSOS]2.0.CO;2.

3. Big Bang Theory

58 **In 2000, the sauropod was reassembled:** Bradley Keoun, "Replica of Dinosaur Fossil Gives O'Hare Passengers Monstrous Welcome," *Chicago Tribune*, January 20, 2000, articles.chicagotribune.com/2000-01-20 /news/0001200303_1_dinosaur-skeleton-brachiosaurus-love-dinosaurs.

59 **That skull—and a few other parts:** Michael P. Taylor, "A ReEvaluation of *Brachiosaurus altithorax* Riggs 1903 (Dinosauria, Sauropod) and Its Generic Separation from *Giraffatitan brancai* (Janensh 1914)," *Journal of Vertebrate Paleontology* 29, no. 3 (2009): 787–806, dx.doi.org/10.1671 /039.029.0309.

60   **As the British paleontologist Derek Ager:** D. U. Ager, *Principles of Paleoecology* (London: McGraw-Hill, 1963).

60   **Henry Fairfield Osborn pointed to affectionate occasions:** Henry Fairfield Osborn, "*Tyrannosaurus*, Upper Cretaceous Carnivorous Dinosaur (Second Communication)," *Bulletin of the American Museum of Natural History* 22, no. 16 (1906): 281–96.

61   **George Murray Levick:** Robin McKie, "'Sexual Depravity' of Penguins that Antarctic Scientist Dared Not Reveal," *Guardian*, June 9, 2012, www.guardian.co.uk/world/2012/jun/09/sex-depravity-penguins-scott -antarctic.

62   **The fossil was embedded in the chest:** Paul E. Fisher et al., "Cardiovascular Evidence for an Intermediate or Higher Metabolic Rate in an Ornithischian Dinosaur," *Science* 288, no. 5465 (2000): 503–505, doi:10.1126 /science.288.5465.503.

62   **But in 2011 another team:** Timothy P. Cleland, Michael K. Stoskopf, and Mary H. Schweitzer, "Histological, Chemical, and Morphological Reexamination of the 'Heart' of a Small Late Cretaceous *Thescelosaurus*," *Naturwissenschaften* 98, no. 3 (2011): 203–11.

62   **a 380-million-year-old armored fish:** Per Ahlberg et al., "Pelvic Claspers Confirm Chondrichthyan-Like Internal Fertilization in Arthrodires," *Nature* 460 (2009): 888–89, doi:10.1038/nature08176.

63   *Scipionyx* **that was found in Italy:** Cristiano Dal Sasso and Marco Signore, "Exceptional Soft Tissue Preservation in a Theropod Dinosaur from Italy," *Nature* 392 (1998): 383–87, doi:10.1038/32884.

63   **Inside the hips:** Tamaki Sato et al., "A Pair of Shelled Eggs Inside a Female Dinosaur," *Science* 308, no. 5720 (2005): 375, doi:10.1126/science .1110578.

65   **According to the ornithologist Kevin McCracken:** Kevin G. McCracken, "The 20-cm Spiny Penis of the Argentine Lake Duck (*Oxyura vittata*)," *The Auk* 117, no. 3 (2000), 820–25.

65   **Other ducks, male and female, are similarly famous:** Patricia L. R. Brennan et al., "Coevolution of Male and Female Genital Morphology in Waterfowl," *PLoS ONE* 2, no. 5 (2007), e418, doi:10.1371/journal .pone.0000418.

65   **As determined by:** P.L.R. Brennan et al., "Independent Evolutionary Reductions of the Phallus in Basal Birds," *Journal of Avian Biology* 39, no. 5 (2008): 487–92, doi:10.1111/j.2008.0908-8857.04610.x.

65   **male alligators, crocodiles, and gharials have penises, too:** Brandon C. Moore, Ketan Mathavan, and Louis J. Guillette Jr., "Morphology and Histochemistry of Juvenile Male American Alligator (*Alligator mississippiensis*) Phallus," *The Anatomical Record: Advances in Integrative Anatomy and Evolutionary Biology* 295 (2012): 328–37, doi:10.1002/ar.21521; Thomas Ziegler and Sven Olbort, "Genital Structures and Sex Identification in Crocodiles," *Crocodile Specialist Group Newsletter* 26, no. 3 (2007): 16–17.

66   **In 2006, Steve Wang:** Steve C. Wang and Peter Dodson, "Estimating the Diversity of Dinosaurs," *Proceedings of the National Academy of Sciences* 103, no. 37 (2006): 13601–605, doi:10.1073/pnas.0606028103.

66   **Peter Dodson highlighted the trouble:** P. Dodson, "Taxonomic Implications of Relative Growth in Lambeosaurine Hadrosaurs," *Systematic Zoology* 24, no. 1 (1975): 37–54.

67   **the Canadian paleontologists David Evans and Robert Reisz:** D. C. Evans and R. R. Reisz, "Anatomy and Relationships of *Lambeosaurus magnicristatus*, a Crested Hadrosaurid Dinosaur (Ornithischia) from the Dinosaur Park Formation, Alberta," *Journal of Vertebrate Paleontology* 27, no. 2 (2007): 373–93.

67   **During the 1990s, some paleontologists:** Kenneth Carpenter, "Variation in *Tyrannosaurus rex*," in *Dinosaur Systematics: Approaches and Perspectives*, ed. Kenneth Carpenter and Philip J. Currie (New York: Cambridge University Press, 1990), 141. P. L. Larson, "*Tyrannosaurus sex*," in *Dino Fest: Proceedings of a Conference for the General Public*, ed. Gary D. Rosenberg and D. L. Wolberg, Paleontological Society Special Publications 7 (Knoxville: The Paleontological Society, 1994), 139.

68   **As paleontologists later found:** Gregory M. Erickson, A. Kristopher Lappin, and Peter Larson, "Androgynous rex—The Utility of Chevrons for Determining the Sex of Crocodilians and Non-Avian Dinosaurs," *Zoology* 108, no. 4 (2005): 277–86, dx.doi.org/10.1016/j.zool.2005.08.001.

68   **no one has yet found an unequivocal case:** Kevin Padian and Jack R. Horner, "The Evolution of 'Bizarre Structures' in Dinosaurs: Biomechanics, Sexual Selection, Social Selection or Species Recognition?," *Journal of Zoology* 283, no. 1 (2011): 3–17, doi:10.1111/j.1469-7998.2010.00719.x.

68   **In 2000, a special specimen of *T. rex*:** Jack Horner and James Gorman, *How to Build a Dinosaur: Extinction Doesn't Have to Be Forever* (New York: Dutton, 2009), 61–67.

69   **The *Tyrannosaurus* had been pregnant:** Mary H. Schweitzer, Jennifer L. Wittmeyer, and John R. Horner, "Gender-Specific Reproductive Tissue in Ratites and *Tyrannosaurus rex*," *Science* 308, no. 5727 (2005): 1456–60, doi:10.1126/science.1112158.

69   **Andrew Lee and Sarah Werning:** A. H. Lee and S. Werning, "Sexual Maturity in Growing Dinosaurs Does Not Fit Reptilian Growth Models," *PNAS* 105, no. 2 (2008): 582–87, doi:10.1073/pnas.0708903105.

71   **In his children's book *The Year of the Dinosaur*:** Edwin H. Colbert, *The Year of the Dinosaur*, illustrated by Margaret Colbert (New York: Charles Scribner's Sons, 1977), 101.

71   **William Service added a little more color:** William Stout and William Service, *The Dinosaurs: A Fantastic View of a Lost Era* (New York: Byron Preiss Books, 1981), 13–14.

72   **the strange adornments:** Scott D. Sampson, "Bizarre Structures and Dinosaur Evolution," in *Dinofest International: Proceedings of a Symposium Held at*

Arizona State University, ed. Donald L. Wolberg, Edmund Stump, and Gary D. Rosenberg (Philadelphia: The Academy of Natural Sciences, 1997), 39–45.

72 **But in 1996 the zoologists:** Robert E. Simmons and Lue Scheepers, "Winning by a Neck: Sexual Selection in the Evolution of Giraffe," *The American Naturalist* 148, no. 5 (1996): 771–86.

73 **Phil Senter applied the same idea to sauropods:** Phil Senter, "Necks for Sex: Sexual Selection as an Explanation for Sauropod Dinosaur Neck Elongation," *Journal of Zoology* 271, no. 1 (2007): 45–53, doi:10.1111 /j.1469-7998.2006.00197.x.

73 **But the sauropod experts:** Michael P. Taylor, Mathew J. Wedel, and Darren Naish, "Head and Neck Posture in Sauropod Dinosaurs Inferred from Extant Animals," *Acta Palaeontologica Polonica* 5, no. 2 (2009): 213–20, doi:10.4202/app.2009.0007; M. P. Taylor et al., "The Long Necks of Sauropods Did Not Evolve Primarily Through Sexual Selection," *Journal of Zoology* 285, no. 2 (2011): 150–61, doi:10.1111/j.1469- 7998.2011 .00824.x.

73 **Giraffes, too, have been shown to gain a feeding advantage:** Elissa Z. Cameron and Johan T. du Toit, "Winning by a Neck: Tall Giraffes Avoid Competing with Shorter Browsers," *The American Naturalist* 169, no. 1 (2007): 130–35; G. Mitchell, S. J. van Sittert, and J. D. Skinner, "Sexual Selection Is Not the Origin of Long Necks in Giraffes," *Journal of Zoology* 278, no. 4 (2009): 281–86, doi:10.1111/j.1469-7998.2009.00573.x; R. E. Simmons and R. Altwegg, "Necks-for-Sex or Competing Browsers? A Critique of Ideas on the Evolution of Giraffe," *Journal of Zoology* 282, no. 1 (2010): 6–12, doi:10.1111/j.1469-7998.2010.00711.x.

75 **The biomechanics expert R. McNeill Alexander:** R. M. Alexander, *Dynamics of Dinosaurs and Other Extinct Giants* (New York: Columbia University Press, 1989), 57–58.

76 **However they did it:** Timothy E. Isles, "The Socio-Sexual Behaviour of Extant Archosaurs: Implications for Understanding Dinosaur Behaviour," *Historical Biology* 21, nos. 3–4 (2009): 139–214.

77 **Among other things, Mallison discovered:** Heinrich Mallison, "CAD Assessment of the Posture and Range of Motion of *Kentrosaurus aethiopicus* Hennig, 1915," *Swiss Journal of Geosciences* 103, no. 2 (2010): 211– 33, doi:10.1007/s00015-010-0024-2; H. Mallison, "Defense Capabilities of *Kentrosaurus aethiopicus* Hennig, 1915," *Palaeontologia Electronica* 14, no. 2 (2011), 10A:25p.

4. The Dinosaurs, They Are a-Changin'

79 **The AMNH paleontologist Mark Norell:** Mark A. Norell et al., "A Theropod Dinosaur Embryo and the Affinities of the Flaming Cliffs Dinosaur Eggs," *Science* 266, no. 5186 (1994): 779–82, doi:10.1126/science .266.5186.779.

81   **Robert Reisz and colleagues found that:** R. R. Reisz et al., "Embryonic Skeletal Anatomy of the Sauropodomorph Dinosaur *Massospondylus* from the Lower Jurassic of South Africa," *Journal of Vertebrate Paleontology* 30, no. 6 (2010): 1653–65, doi:10.1080/02724634.2010.521604

81   **And there were more eggs:** R. R. Reisz et al., "Oldest Known Dinosaurian Nesting Site and Reproductive Biology of the Early Jurassic Sauropodomorph *Massospondylus*," *PNAS* 109, no. 7 (2012): 2428–33, doi:10.1073/pnas.1109385109.

81   **One of the most fantastic dinosaur nesting grounds:** Lucas E. Fiorelli et al., "The Geology and Palaeoecology of the Newly Discovered Cretaceous Neosauropod Hydrothermal Nesting Site in Sanagasta (Los Llanos Formation), La Rioja, Northwest Argentina," *Cretaceous Research* 36 (2011): 94–117, dx.doi.org/10.1016/j.cretres.2011.12.002.

82   **paleontologists have found burrows:** David J. Varricchio, Anthony J. Martin, and Yoshihiro Katsura, "First Trace and Body Fossil Evidence of a Burrowing, Denning Dinosaur," *Proceedings of the Royal Society B* 274, no. 1616 (2007): 1361–68, doi:10.1098/rspb.2006.0443.

82   **According to David Varricchio:** D. J. Varricchio, "A Distinct Dinosaur Life History?," *Historical Biology* 23, no. 1 (2011): 91–107, doi:10.1080/08912963.2010.500379.000.

83   **It all started in July of that year:** John B. Scannella and John R. Horner, "*Torosaurus* Marsh, 1891, is *Triceratops* Marsh, 1889 (Ceratopsidae: Chasmosaurinae): Synonymy Through Ontogeny," *Journal of Vertebrate Paleontology* 30, no. 4 (2010): 1157–68, dx.doi.org/10.1080/02724634.2010.483632.

84   **discovered its remains in 1887:** Kenneth Carpenter, "'Bison' alticornis and O.C. Marsh's Early Views on Ceratopsians," in *Horns and Beaks: Ceratopsian and Ornithopod Dinosaurs*, ed. K. Carpenter (Bloomington: Indiana University Press, 2007), 349.

87   **Mark Goodwin, the University of California, Berkeley, paleontologist:** John R. Horner and Mark B. Goodwin, "Major Cranial Changes During *Triceratops* Ontogeny," *Proceedings of the Royal Society B* 273, no. 1602 (2006): 2757–61, doi:10.1098/rspb.2006.3643.

88   **During the 1980s and 1990s:** Catherine A. Forster, "Species Resolution in *Triceratops*: Cladistic and Morphometric Approaches," *Journal of Vertebrate Paleontology* 16, no. 2 (1996): 259–70; John H. Ostrom and Peter Wellnhofer, "The Munich Specimen of *Triceratops* with a Revision of the Genus," *Zitteliana* 14 (1986): 111–58.

89   **It wasn't until 2006:** Mark B. Goodwin et al., "The Smallest Known *Triceratops* Skull: New Observations on Ceratopsid Cranial Anatomy and Ontogeny," *Journal of Vertebrate Paleontology* 26, no. 1 (2006): 103–12.

89   **Indeed, a different dinosaur:** John B. Scannella and John R. Horner, "'Nedoceratops': An Example of a Transitional Morphology," *PLoS ONE* 6, no. 12 (2011): e28705, doi:10.1371/journal.pone.

90   **"My third horn":** Limericks, *Wait Wait . . . Don't Tell Me!*, NPR, August 7, 2010. www.npr.org/templates/story/story.php?storyId=129039425.

91   **What's more, Farke has noted:** Andrew A. Farke, "Anatomy and Taxonomic Status of the Chasmosaurine Ceratopsid *Nedoceratops hatcheri* from the Upper Cretaceous Lance Formation of Wyoming, U.S.A," *PLoS ONE* 6, no. 1 (2011): e16196, doi:10.1371/journal.pone.0016196; Nicholas R. Longrich and Daniel J. Field, "Torosaurus Is Not Triceratops: Ontogeny in Chasmosaurine Ceratopsids as a Case Study in Dinosaur Taxonomy," *PLoS ONE* 7, no. 2 (2012): e32623, doi:10.1371/journal.pone .0032623.

92   **when Peter Dodson surveyed:** P. Dodson, "Taxonomic Implications of Relative Growth in Lambeosaurine Hadrosaurs," *Systematic Zoology* 24, no. 1 (1975): 37–54, www.jstor.org/stable/2412696.

92   **The same was true of *"Brachyceratops":*** Charles W. Gilmore, "*Brachyceratops*, a Ceratopsian Dinosaur from the Two Medicine Formation of Montana, with Notes on Associated Fossil Reptiles," *United States Geological Survey Professional Paper* 103 (1917); Andrew T. McDonald, "A Subadult Specimen of *Rubeosaurus ovatus* (Dinosauria: Ceratopsidae), with Observations on Other Ceratopsids from the Two Medicine Formation," *PLoS ONE* 6, no. 8 (2011): e22710, doi:10.1371/journal.pone.0022710.

93   **But according to Nicolás Campione:** Nicolás E. Campione and David C. Evans, "Cranial Growth and Variation in Edmontosaurs (Dinosauria: Hadrosauridae): Implications for Latest Cretaceous Megaherbivore Diversity in North America," *PLoS ONE* 6, no. 9 (2011), e25186, doi:10.1371/journal.pone.0025186.

93   **In 1946, Charles Gilmore:** C. W. Gilmore, "A New Carnivorous Dinosaur from the Lance Formation of Montana," *Smithsonian Miscellaneous Collections* 106 (1946): 1–19.

93   **paleontologists regularly mistook young tyrannosaurs:** Takanobu Tsuihiji et al., "Cranial Osteology of a Juvenile Specimen of *Tarbosaurus bataar* (Theropoda, Tyrannosauridae) from the Nemegt Formation (Upper Cretaceous) of Bugin Tsav, Mongolia," *Journal of Vertebrate Paleontology* 31, no. 3 (2011): 497–517.

94   **As Thomas Carr:** Thomas D. Carr, "Craniofacial Ontogeny in Tyrannosauridae (Dinosauria, Coelurosauria)," *Journal of Vertebrate Paleontology* 19, no. 3 (1999): 497–520, www.jstor.org/stable/4524012; T. D. Carr and Thomas E. Williamson, "Diversity of Late Maastrichtian Tyrannosauridae (Dinosauria: Theropoda) from Western North America," *Zoological Journal of the Linnean Society* 142, no. 4 (2004): 479–523; Lawrence M. Witmer and Ryan C. Ridgely, "The Cleveland Tyrannosaur Skull (*Nanotyrannus* or *Tyrannosaurus*): New Findings Based on CT Scanning, with Special Reference to the Braincase," *Kirtlandia* 57 (2010): 61–81; Denver W. Fowler et al., "Reanalysis of '*Raptorex kriegsteini*': A Juvenile Tyrannosaurid Dino-

saur from Mongolia," *PLoS One* 6, no. 6 (2011): e21376, doi:10.1371/journal
.pone.0021376.

5. Jurassic Thunder

100 Brachiosaurus was as big as dinosaurs got: Elmer S. Riggs,
"*Brachiosaurus altithorax*, the Largest Known Dinosaur," *American Journal of
Science* (Series 4) 15, no. 88 (1903): 299–306, doi:10.2475/ajs.s4-15.88.299;
Ruth E. Moore, *Evolution*, Young Readers Nature Library (Alexandria,
VA: Time-Life Books, 1979), 94–95.

104 As Matt Wedel remarked: Mathew J. Wedel, "A Monument of Inefficiency: The Presumed Course of the Recurrent Laryngeal Nerve in Sauropod Dinosaurs," *Acta Palaeontologica Polonica* 57, no. 2 (2012): 251–56,
dx.doi.org/10.4202/app.2011.0019.

104 As the paleontologist Stephen Jay Gould: Stephen J. Gould, "The
Panda's Thumb," in *The Panda's Thumb: More Reflections in Natural History*
(New York: W. W. Norton, 1980), 19.

108 The paleontologist Edwin Colbert and his collaborators: Edwin H. Colbert, Raymond B. Cowles, and Charles M. Bogert, "Temperature Tolerances in the American Alligator and Their Bearing on the Habits,
Evolution, and Extinction of the Dinosaurs," *Bulletin of the American Museum
of Natural History* 86 (1946): 327–74.

109 a landmark study of mammal bones: Meike Köhler et al., "Seasonal Bone Growth and Physiology in Endotherms Shed Light on Dinosaur Physiology," *Nature* 487 (2012): 358–61, doi:10.1038/nature11264.

109 As the paleontologist Kevin Padian remarked: Kevin Padian,
"Evolutionary Physiology: A Bone for All Seasons," *Nature* 487 (2012):
310–11, doi: 10.1038/nature11382.

110 As the paleontologist Stephen Brusatte wrote: S. L. Brusatte,
Dinosaur Paleobiology (Chichester, UK: Wiley-Blackwell, 2012), 216–26.

111 the physiology and biology: P. Martin Sander et al., "Biology of the
Sauropod Dinosaurs: The Evolution of Gigantism," *Biological Reviews* 86,
no. 1 (2011): 117–55.

112 paleontologists Christine Janis and Matthew Carrano: Christine M. Janis and Matthew Carrano, "Scaling of Reproductive Turnover
in Archosaurs and Mammals: Why Are Large Terrestrial Mammals So
Rare?," *Annales Zoologici Fennici* 28 (1992): 201–16; Jan Werner and Eva
Maria Griebeler, "Reproductive Biology and Its Impact on Body Size:
Comparative Analysis of Mammalian, Avian and Dinosaurian Reproduction," *PLoS ONE* 6, no. 12 (2011), e28442, doi:10.1371/journal.pone
.0028442.

113 The skull of a juvenile *Diplodocus*: John A. Whitlock, Jeffrey A.
Wilson, and Matthew C. Lamanna, "Description of a Nearly Complete

Juvenile Skull of *Diplodocus* (Sauropoda: Diplodocoidea) from the Late Jurassic of North America," *Journal of Vertebrate Paleontology* 30, no. 2 (2010): 442–57, doi:10.1080/02724631003617647.

6. Dinosaur Society

121 **the site where *Deinonychus*:** John H. Ostrom, "Osteology of *Deinonychus antirrhopus*, an Unusual Theropod from the Lower Cretaceous of Montana," *Bulletin of the Peabody Museum of Natural History* 30 (1969); W. Desmond Maxwell and J. H. Ostrom, "Taphonomy and Paleobiological Implications of *Tenontosaurus-Deinonychus* Associations," *Journal of Vertebrate Paleontology* 15, no. 4 (1995): 707–12.

122 **Yet the quarry Ostrom described:** Brian T. Roach and Daniel L. Brinkman, "A Reevaluation of Cooperative Pack Hunting and Gregariousness in *Deinonychus antirrhopus* and Other Nonavian Theropod Dinosaurs," *Bulletin of the Peabody Museum of Natural History* 48, no. 1 (2007): 103–38, dx.doi.org/10.3374/0079-032X(2007)48[103:AROCPH]2.0.CO;2.

122 **This site, tucked away:** Michael J. Ryan et al., "The Taphonomy of a *Centrosaurus* (Ornithischia: Cer[a]topsidae) Bone Bed from the Dinosaur Park Formation (Upper Campanian), Alberta, Canada, with Comments on Cranial Ontogeny," *PALAIOS* 16 (2001): 482–506.

125 **The paleontologist Roland T. Bird:** See R. T. Bird, *Bones for Barnum Brown: Adventures of a Dinosaur Hunter* (Fort Worth: Texas Christian University Press, 1985).

128 **One trackway made by these dinosaurs:** Rihui Li et al., "Behavioral and Faunal Implications of Early Cretaceous Deinonychosaur Trackways from China," *Naturwissenschaften* 95, no. 3 (2008): 185–91, doi:10.1007/s00114-007-0310-7; Alexander Mudroch et al., "Didactyl Tracks of Paravian Theropods (Maniraptora) from the Middle Jurassic of Africa," *PLoS ONE* 6, no. 2 (2011), e14642, doi:10.1371/journal.pone.0014642.

129 **In fact, Buckland proclaimed:** William Buckland, *Geology and Mineralogy Considered with Reference to Natural Theology* (London: William Pickering, 1837).

130 **The fossil wasn't actually:** Kenneth Carpenter et al., "Evidence for Predator-Prey Relationships: Examples for *Allosaurus* and *Stegosaurus*," in *The Carnivorous Dinosaurs*, ed. K. Carpenter (Bloomington: Indiana University Press, 2005), 325.

131 **A few years before, Farke had:** Andrew A. Farke, "Horn Use in Triceratops (Dinosauria: Ceratopsidae): Testing Behavioral Hypotheses Using Scale Models," *Palaeontologia Electronica* (2004).

131 **When the Yale University paleontologist Richard Swann Lull:** Richard S. Lull, "Restoration of the Horned Dinosaur *Diceratops*," *The American Journal of Science* 4, 4 (1905): 420–22.

132 **If the dinosaurs were fighting:** A. A. Farke, Ewan D. S. Wolff, and Darren H. Tanke, "Evidence of Combat in *Triceratops*," *PLoS ONE* 4, no. 1 (2009), e4252, doi:10.1371/journal.pone.0004252.

134 **Goodwin's work with Jack Horner:** Mark B. Goodwin and John R. Horner, "Cranial Histology of Pachycephalosaurs (Ornithischia: Marginocephalia) Reveals Transitory Structures Inconsistent with Head-Butting Behavior," *Paleobiology* 30, no. 2 (2004): 253–67, doi:10.1666/0094-8373 (2004)030<0253:CHOPOM>2.0.CO;2; J. R. Horner and M. B. Goodwin, "Extreme Cranial Ontogeny in the Upper Cretaceous Dinosaur *Pachycephalosaurus*," *PLoS ONE* 4, no. 10 (2009): e7626, doi:10.1371/journal.pone.0007626.

134 **But in 2012, Joseph Peterson:** J. E. Peterson and Christopher P. Vittore, "Cranial Pathologies in a Specimen of *Pachycephalosaurus*," *PLoS ONE* 7, no. 4 (2012): e36227, doi:10.1371/journal.pone.0036227.

7. Dinosaur Feathers

139 **The paleoartist Gregory S. Paul:** G. S. Paul, *Predatory Dinosaurs of the World: A Complete Illustrated Guide* (New York; Simon & Schuster, 1988), 126–27.

142 **As the paleontologist Hugh Falconer:** H. Falconer, Letter to Darwin, January 3, 1863, Darwin Correspondence Database, www.darwinproject .ac.uk/entry-3899, accessed July 13, 2012.

142 **who obtained the first skeletal specimen:** Richard Owen, "On the *Archaeopteryx* of von Meyer, with a Description of the Fossil Remains of a Long-Tailed Species, from the Lithographic Stone of Solenhofen," *Philosophical Transactions of the Royal Society of London* 153 (1863): 33–47.

145 **It turns out that no one knows:** Peter Wellnhofer, *Archaeopteryx: The Icon of Evolution*, rev. English ed., trans. Frank Haase (Munich: Verlag Dr. Friedrich Pfeil, 2009).

149 **included a restoration of the Triassic dinosaur:** Robert T. Bakker, "Dinosaur Renaissance," in *The Scientific American Book of Dinosaurs*, ed. Gregory S. Paul (New York: St. Martin's Griffin, 2003), 331–44.

150 **John Ostrom, who was chiefly responsible:** Malcolm Browne, "Feathery Fossil Hints Dinosaur-Bird Link," *New York Times*, October 19, 1996, www.nytimes.com/1996/10/19/us/feathery-fossil-hints-dinosaur -bird-link.html.

152 **Thanks to these finds:** Xing Xu et al., "A Gigantic Feathered Dinosaur from the Lower Cretaceous of China," *Nature* 484 (2012): 92–95, doi:10.1038/nature10906.

153 **Two dinosaurs—each about as far removed:** Lawrence L. Witmer, "Dinosaurs: Fuzzy Origins for Feathers," *Nature* 458 (2009): 293–95.

153 **a fuzzy juvenile dinosaur:** Oliver W. M. Rauhut et al., "Exceptionally

Preserved Juvenile Megalosauroid Theropod Dinosaur with Filamentous Integument from the Late Jurassic of Germany," *PNAS* 109, no. 29 (2012): 11746–51, doi:10.1073/pnas.1203238109.

155 **As Charles Darwin wrote:** C. R. Darwin, *The Descent of Man, and Selection in Relation to Sex*, vol. 1 (London: John Murray, 1871), 3.

156 **The *Archaeopteryx* feather was black:** Ryan M. Carney et al., "New Evidence on the Colour and Nature of the Isolated *Archaeopteryx* Feather," *Nature Communications* 3, article no. 637 (2012), doi:10.1038/ncomms1642.

157 **They did just that:** Jakob Vinther et al., "The Colour of Fossil Feathers," *Biology Letters* 4, no. 5 (2008): 522–25, doi:10.1098/rsbl.2008.0302.

158 **the following year, Vinther led:** J. Vinther et al., "Structural Coloration in a Fossil Feather," *Biology Letters* 6, no. 1 (2009): 128–31, doi:10.1098/rsbl.2009.0524.

158 **On January 27, 2010:** Fucheng Zhang et al., "Fossilized Melanosomes and the Colour of Cretaceous Dinosaurs and Birds," *Nature* 463 (2010): 1075–78.

158 **Vinther's team countered:** Quanguo Li et al., "Plumage Color Patterns of an Extinct Dinosaur," *Science* 327, no. 5971 (2010): 1369–72, doi:10.1126/science.1186290.

159 **the feathered dinosaur *Microraptor*:** Quanguo Li et al., "Reconstruction of *Microraptor* and the Evolution of Iridescent Plumage," *Science* 335, no. 6073 (2012): 1215–19, doi:10.1126/science.1213780.

8. Hadrosaur Harmonics and Tyrannosaur Tastes

162 **The Fayetteville State University paleontologist:** Phil Senter, "Voices of the Past: A Review of Paleozoic and Mesozoic Animal Sounds," *Historical Biology* 20, no. 4 (2008): 255–87, doi:10.1080/08912960903033327.

167 **the Canadian paleontologist William Parks:** W. A. Parks, "*Parasaurolophus walkeri*, a New Genus and Species of Crested Trachodont Dinosaur," *University of Toronto Studies, Geology Series* 13 (1922): 1–32.

168 **Henry Fairfield Osborn described an exceptional *Edmontosaurus*:** H. F. Osborn, "Integument of the Iguanodont Dinosaur *Trachodon*," *Memoirs of the AMNH* (new series) 1, no. 2 (1912): 33–54.

169 **Even John Ostrom:** J. H. Ostrom, "The Cranial Crests of Hadrosaurian Dinosaurs," *Yale Peabody Museum of Natural History Postilla* 62 (1962): 1–29.

169 **When James Hopson:** J. A. Hopson, "The Evolution of Cranial Display Structures in Hadrosaurian Dinosaurs," *Paleobiology* 1, no. 1 (1975): 21–43, www.jstor.org/stable/2400327.

170 **David Weishampel used an improvised model:** D. B. Weishampel, "Acoustic Analyses of Potential Vocalization in Lambeosaurine Dinosaurs (Reptilia: Ornithischia)," *Paleobiology* 7, no. 2 (1981): 252–61; Weishampel,

"Dinosaurian Cacophony: Inferring Function in Extinct Animals," *Bio-Science* 47, no. 3 (1997): 150–59, www.jstor.org/stable/1313034.

172 **David Evans is one of the paleontologists:** D. C. Evans, "Nasal Cavity Homologies and Cranial Crest Function in Lambeosaurine Dinosaurs," *Paleobiology* 32, no. 1 (2006): 109–25, dx.doi.org/10.1666/04027.1; D. C. Evans, Ryan C. Ridgely, and L. M. Witmer, "Endocranial Anatomy of Lambeosaurine Hadrosaurids (Dinosauria: Ornithischia): A Sensorineural Perspective on Cranial Crest Function," *The Anatomical Record*, 292 (2009): 1315–37.

173 **In 2005, Otto Gleich:** O. Gleich, Robert J. Dooling, and Geoffrey A. Manley, "Audiogram, Body Mass, and Basilar Papilla Length: Correlations in Birds and Predictions for Extinct Archosaurs," *Naturwissenschaften* 92, no. 12 (2005): 595–98, doi:10.1007/s00114-005-0050-5.

175 **At 1994's Dino Fest:** J. R. Horner, "Steak Knives, Beady Eyes, and Tiny Little Arms (A Portrait of *T. rex* as a Scavenger)," in *Dino Fest: Proceedings of a Conference for the General Public*, ed. Gary D. Rosenberg and D. L. Wolberg, *Paleontological Society Special Publications* 7 (Knoxvillle: The Paleontological Society, 1994), 157.

176 ***Tyrannosaurus* was certainly capable:** Thomas R. Holtz Jr., "A Critical Re-Appraisal of the Obligate Scavenging Hypothesis for *Tyrannosaurus rex* and Other Tyrant Dinosaurs," in *Tyrannosaurus rex: The Tyrant King*, ed. Peter L. Larson and Kenneth Carpenter (Bloomington: Indiana University Press, 2008), 371.

9. In the Bones

178 **Chris Brochu's exhaustive monograph:** Christopher A. Brochu, "Osteology of *Tyrannosaurus Rex:* Insights from a Nearly Complete Skeleton and High-Resolution Computed Tomographic Analysis of the Skull," *Journal of Vertebrate Paleontology* 22, sup4 (2003), doi:10.1080/02724634.2003.10010947.

179 **Shortly after Sue was discovered:** Peter Larson and Kristin Donnan, *Rex Appeal: The Amazing Story of Sue, the Dinosaur That Changed Science, the Law, and My Life* (Montpelier, VT: Invisible Cities Press, 2002), 1–2.

179 **predatory dinosaurs often fought:** Darren H. Tanke and Philip J. Currie, "Head-Biting Behavior in Theropod Dinosaurs: Paleopathological Evidence," *GAIA* 15 (1998): 167–84.

180 **The veterinarian Ewan Wolff:** E.D.S. Wolff et al., "Common Avian Infection Plagued the Tyrant Dinosaurs," *PLoS ONE* 4, no. 9 (2009): e7288, doi:10.1371/journal.pone.0007288.

181 **a young *T. rex* nicknamed "Jane":** Joseph E. Peterson et al., "Face Biting on a Juvenile Tyrannosaurid and Behavioral Implications," *PALAIOS* 24, no. 11 (2009): 780–84, doi: 10.2110/palo.2009.p09-056r.

181 **Coarsely serrated teeth:** William L. Abler, "The Serrated Teeth of Tyrannosaurid Dinosaurs, and Biting Structures in Other Animals," *Paleobiology* 18, no. 2 (1992): 161–83, www.jstor.org/stable/2400997.

181 **Cannibalism is another possibility:** Nicholas R. Longrich et al., "Cannibalism in *Tyrannosaurus rex*," *PLoS ONE* 5, no. 10 (2010): e13419, doi:10.1371/journal.pone.0013419.

182 **author Brian Aldiss:** B. Aldiss, "Poor Little Warrior!," in *Behold the Mighty Dinosaur*, ed. David Jablonski (New York: Elsevier/Nelson Books, 1981), 180.

183 **Tyrannosaur coprolites:** Karen Chin et al., "A King-Sized Theropod Coprolite," *Nature* 393 (1998): 680–82; K. Chin et al., "Remarkable Preservation of Undigested Muscle Tissue Within a Late Cretaceous Tyrannosaurid Coprolite from Alberta, Canada," *PALAIOS* 18, no. 3 (2003): 286–94, www.jstor.org/stable/3515739.

183 **the paleontologists George Poinar and Arthur Boucot:** G. Poinar Jr. and A. J. Boucot, "Evidence of Intestinal Parasites of Dinosaurs," *Parasitology* 133, no. 2 (2006), 245–49, doi:10.1017/S0031182006000138.

184 **huge, 165-million-year-old fleas:** Diying Huang et al., "Diverse Transitional Giant Fleas from the Mesozoic Era of China," *Nature* 483 (2012): 201–04, doi:10.1038/nature10839.

185 **Organized by Tanke:** D. H. Tanke and Bruce M. Rothschild, *Dinosores: An Annotated Bibliography of Dinosaur Paleopathology and Related Topics— 1838–2001*, New Mexico Museum of Natural History and Science Bulletin no. 20, 2002.

185 **most of them in hadrosaurs:** B. M. Rothschild et al., "Epidemiologic Study of Tumors in Dinosaurs," *Naturwissenschaften* 90, no. 11 (2003): 495– 500, doi:10.1007/s00114-003-0473-9.

186 **pathologist Roy L. Moodie:** R. L. Moodie, *Studies in Paleopathology* (New York: Paul B. Hoeber, 1918); L. C. Natarajan et al., "Bone Cancer Rates in Dinosaurs Compared with Modern Vertebrates," *Transactions of the Kansas Academy of Science* 110, nos. 3–4 (2007): 155–58.

187 **One dinosaur that I feel especially sorry for:** Brent H. Breithaupt, "The Discovery of a Nearly Complete *Allosaurus* from the Jurassic Morrison Formation, Eastern Bighorn Basin, Wyoming," in *Resources of the Bighorn Basin: Forty-seventh Annual Field Conference Guidebook*, ed. C. E. Bowen, S. C. Kirkwood, and T. S. Miller (Casper: Wyoming Geological Association, 1996), 309.

187 **paleontologist Rebecca Hanna cataloged:** R. C. Hanna, "Multiple Injury and Infection in a Sub-Adult Theropod Dinosaur *Allosaurus fragilis* with Comparisons to Allosaur Pathology in the Cleveland-Lloyd Dinosaur Quarry Collection," *Journal of Vertebrate Paleontology* 22, no. 1 (2002): 76–90.

188 **this injured dinosaur wasn't alone:** Martin G. Lockley et al., "Limping Dinosaurs? Trackway Evidence for Abnormal Gaits," *Ichnos* 3 (1994): 193–202.

10. Dinosaurs Undone

197 **As Michael Benton has emphasized:** M. J. Benton, "Scientific Methodologies in Collision: The History of the Study of the Extinction of Dinosaurs," *Evolutionary Biology* 24 (1990): 371–400.

197 **dinosaurs "died a natural death":** R. S. Lull, *Organic Evolution: A Text-book* (New York: The MacMillan Company, 1917), 220–25.

198 **one paleontologist who didn't follow this trend:** G. R. Wieland, "Dinosaur Extinction," *American Naturalist* 59, no. 665 (1925): 557–65.

199 **entomologist Stanley Flanders:** S. E. Flanders, "Did the Caterpillar Exterminate the Giant Reptile?," *Journal of Research on the Lepidoptera* 1, no. 1 (1962): 85–88.

202 **Instead, in a 1980 *Science* paper:** Luis W. Alvarez et al., "Extraterrestrial Cause for the Cretaceous-Tertiary Extinction," *Science* 208, no. 4448 (1980): 1095–1108.

205 **a *New York Times* reporter:** Malcolm W. Browne, "Dinosaur Experts Resist Meteor Extinction Idea," *New York Times*, October 29, 1985, www.nytimes.com/1985/10/29/science/dinosaur-experts-resist-meteor-extinction-idea.html?pagewanted=all.

205 **"The arrogance of those people":** M. B. Browne, "Dinosaur Experts Resist Meteor Extinction Idea," *New York Times*, October 29, 1985, www.nytimes.com/1985/10/29/science/dinosaur-experts-resist-meteor-extinction-idea.html.

206 **paleontologist David Raup:** D. M. Raup, "The Extinction Debates: A View from the Trenches," in *The Mass-Extinction Debates: How Science Works in a Crisis*, ed. William Glen (Stanford, CA: Stanford University Press, 1994), 145.

207 **geologist Alan Hildebrand:** A. R. Hildebrand et al., "Chicxulub Crater: A Possible Cretaceous/Tertiary Boundary Impact Crater on the Yucatan Peninsula, Mexico," *Geology* 19, no. 9 (1991): 867–71, doi:10.1130/0091-7613(1991)019<0867:CCAPCT>2.3.CO;2.

207 **"biological failings":** D. M. Raup, "The Extinction Debates: A View from the Trenches," in *The Mass-Extinction Debates: How Science Works in a Crisis*, ed. William Glen (Stanford, CA: Stanford University Press, 1994), 150.

209 **a 2010 *Science* paper:** Peter Schulte et al., "The Chicxulub Asteroid Impact and Mass Extinction at the Cretaceous-Paleogene Boundary," *Science* 327, no. 5970 (2010): 1214–18; J. D. Archibald et al., "Cretaceous Extinctions: Multiple Causes," *Science* 328, no. 5981 (2010): 973.

Acknowledgments

A book is like a dinosaur skeleton. Just as a reconstructed dinosaur seldom comes together by the hands of a single person—a varied crew of multiple volunteers and specialists is often involved in discovering, excavating, preparing, and studying any given dinosaur—a book such as this requires the kindness, patience, and help of many.

I owe an immense debt to the friends and professionals who provided me with the encouragement and opportunities to become an author. I'm frustrated that I cannot remember everyone who has helped me, and I apologize to those I have ultimately forgotten by name, but this book could not have existed without the caring attention of numerous friends and colleagues.

Friend and ace science writer Ed Yong introduced me to my enthusiastic agent, Peter Tallack, and I am humbled by Ed and Peter's unflagging support. Peter brought my first book—*Written in Stone*—to the editor Erika Goldman at Bellevue Literary Press, and I am grateful that Erika and the rest of the team at Bellevue took a chance on my debut.

The critical success of the first book, and Peter's tenacious support of what would become *My Beloved Brontosaurus*, caught the attention of this title's editor, Amanda Moon. I met Amanda during the 2010 Science Writers conference in New Haven, Connecticut, and she was enthusiastic about my dinosaur dreams from the very start. Translating the initial idea into an actual book was sometimes a difficult process, especially when my enthusiasm for scientific minutiae threatened to derail the book's narrative, but Amanda is a kind and patient editor who persistently pushed me to keep improving the manuscript. Amanda challenged me to become a better writer and storyteller, and I should note that any hiccups in the story are a result of my own stubbornness. Christoper Richards also contributed suggestions on many chapters, and the copy editor Annie Gottlieb did a heroic job cleaning up the manuscript.

I am grateful for the many friends and experts who supported this project, answered my questions, and let me join them in the field. First and foremost, the paleontology staff at the Natural History Museum of Utah have welcomed me since I first arrived in the Beehive State. Randall Irmis, Mike Getty, Mark Loewen, Carrie Levitt, Jelle Wiersma, Katherine Clayton, Josh Lively, and Eric Lund all contributed to this book in different ways, simply through conversation, in interviews, and by letting me become a field and lab volunteer. The Utah paleontologists Jim Kirkland and Don DeBlieux were very generous with their time and expertise, in addition to the avocational dinosaur fans of Utah Friends of Paleontology.

While writing this book, I was inspired by the time I spent volunteering with field crews led by the paleontologists Jason Schein, Thomas Carr, Scott Williams, and Louis Chiappe. Additionally, this project wouldn't have been the same without discussions and input from Andrew Farke, Sterling Nesbitt, Heinrich Mallison, Tony Martin, Alan Turner, Jerry Harris, Sarah Werning, Bill Parker, Thomas Holtz, Jr., Jakob Vinther, Kirk Johnson, David Fastovsky, Anne Weil, Adrienne Mayor, Joe Sertich, Scott Sampson, Kevin Padian, Mark Goodwin, Jack Horner, Alan Turner, Mark Norrell, Adam Pritchard, David Varrichio, Matt Wedel, Darren Naish, Shaena Montanari, Colleen Farmer, Don Prothero, Carl Mehling, and Ken Lacovara.

I am fortunate to have received guidance from the science writing community, including Carl Zimmer, Bora Zivkovic, Deborah Blum, Thomas Levenson, Maryn McKenna, David Dobbs, Steve Silberman, Jennifer Ouellette, Alok Jha, Adam Rutherford, Mark Henderson, Dave Mosher, Betsy Mason, Virginia Hughes, Brian Wolly, and Kate Wong. I owe a great deal to Laura Helmuth, who edited my blog *Dinosaur Tracking* for Smithsonian for almost four years. I worked with Laura on a near-daily basis, and she is the warmest, most positive editor I have worked with. Under her careful eye, I went from an amateur blogger to a professional science writer, and I feel incredibly fortunate that we worked so closely together for so long—especially since she took such great joy in letting me geek out about dinosaurs each day.

For art, I am thankful for the generosity of Jeff Martz, Niroot Himmapan, Mike Keesey, Robert Walters, Tess Kissinger, Scott Hartman, and Mike Jacobsen. Working with Mark Stutzman on the cover design was an absolute pleasure.

Of course, I will always remember my parents' support of my early dinomania. This book is a tribute to the dreams they encouraged me to develop during my childhood.

Many friends cheered me on during the writing process, but two deserve special mention. Since we met at a ScienceOnline conference several years ago, Scicurious has been an indefatigable friend who has never let me get away with my habit of belittling my own work. She has constantly urged me on, and a writer could not ask for a better confidante. The ever-expanding science writing ecosystem also let me foster a close friendship with Miriam Goldstein, a kindred scientist–science writer hybrid who has encouraged me during especially difficult stretches in this book's composition, not to mention participating in music exchanges that kept my work sound track livelier than usual.

But most of all, I am thankful to my wife, Tracey. She has maintained faith in my abilities even on days when I feel defeated and demoralized, and she has always pushed me to pursue my dreams, even when I took the risk of quitting my job to become a professional freelancer. I could not ask for a better partner and companion as we explore our new home among Utah's wondrous badlands. Every article pitch and book proposal is a struggle, and I value Tracey's insight and support above all others. I am exceptionally lucky to be with such a brilliant and warm spouse. If you enjoyed this book, you have her to thank.

Index

Page numbers in *italics* refer to illustrations.